For the members of my
Spring '91 Chemistry 106 class:

Kevin and the Ladies

Organic and Biochemistry Experiments for the Allied Health Student

Contents

Organic and Biochemistry Experiments for the Allied Health Student

A Prescription for Common Laboratory Ailments

Cynthia Graham Brittain

Kennesaw State College

KENDALL/HUNT PUBLISHING COMPANY
4050 Westmark Drive Dubuque, Iowa 52002

This edition has been printed directly from camera-ready copy.

ISBN 0-8403-5750-8

Printed in the United States of America
10 9 8 7

Preface

This laboratory manual was designed for use in a one quarter organic and biochemistry lecture/laboratory course for students of the allied health sciences.

The eight laboratory experiments included here are fundamental and timeless; each has been carefully selected to illustrate key concepts that are discussed in the lecture portion of the course.

A special feature of this manual is the introductory portion: "A Prescription for Common Laboratory Ailments." This section acknowledges and addresses the fear of chemistry laboratory that is characteristic of many freshman students. The discussion intends to aid students in *anticipating and preparing,* both intellectually and emotionally, for their laboratory experience. The "Pre-Laboratory Prep Sheet" is recommended as a weekly assignment that will serve as the student's *ticket* into the laboratory to perform an experiment. This introduction also emphasizes that *laboratory safety is social as well as personal,* and includes a general list of laboratory safety regulations. The "prescription" closes with a description of the care and usage of common laboratory equipment: burets and graduated cylinders, gravity and vacuum filtration apparatus, and the Bunsen burner.

It must be acknowledged that several of the activities in this manual are re-workings of experiments or exercises that were originally written and/or rewritten by Kennesaw State College chemistry faculty Linda C. Hodges, Patricia H. Reggio, Marianne Scales, and Vera B. Zalkow.

Thank you, friends, for these and other essential contributions to this work.

A Prescription for Common Laboratory Ailments

Why do you need Chemistry Laboratory?

As a student of the allied health sciences, you may question the need to participate in chemistry laboratory. You have chosen a career in the *health* sciences, not in chemistry. Why are you required to perform chemistry experiments and write laboratory reports? Why isn't it enough to learn about chemistry by attending lecture, reading the textbook, and solving problems? There are many reasons why a laboratory experience is an essential part of learning chemistry:

- Chemistry is an *experimental* science. The discoveries that you make in laboratory will help you to appreciate that the information contained in your textbook is the result of inspired reasoning and extensive, careful experimentation.

- Chemistry will literally come alive for you in the laboratory. Your textbook is filledwith graphs, drawings, and photographs; these illustrations are intended to aid in your understanding of the words and symbols used in the text and in the lecture. The laboratory is a living, moving, three-dimensional illustration of chemistry and you are at the center of it. In laboratory, you have the opportunity to see, hear, and perhaps even touch and smell *chemistry in action.* A lecture concept that seemed beyond your grasp can become self-evident when illustated in laboratory.

- Experimentation in laboratory will help you to develop skills in making measurements and observations, and also in communicating your findings with others.

- Laboratory is an opportunity to interact closely with other students in the course. You'll share in one another's successes and frustrations, learning together from each shared experience and developing a cameraderie that will continue outside the laboratory (and that may be a crucial to your success in the course).

Anticipate and Prepare for your Laboratory Experience

Students are rarely indifferent to working in a chemistry laboratory; their emotional reactions to the laboratory experience are as varied as the chemical reactions that they are to study.

- Some students look forward to each session in the laboratory, eagerly anticipating the challenges they must meet and the discoveries they will make. These students enjoy performing the experiment; they think as they work, make predictions of their results, recognize the problems that arise and suggest solutions.

- Other students dread the laboratory experience and approach each session with mounting anxiety. These students timidly, even fearfully, work through the experiment; they don't think about the significance of the procedure, and don't attempt to understand or interpret their results while still in the laboratory. When problems arise, they feel helpless and frustrated.

These reactions are not the result of a difference in the students' knowledge or skills, but of a difference in the way the students have dealt with their anticipation of the laboratory experience. Both types of students anticipate that they will encounter challenges, difficulties, and uncertainties in laboratory. The eager student has dealt with this anticipation by preparing both emotionally and intellectually for the experience. The anxious student has not dealt as effectively with the anticipation. The student has not prepared emotionally for laboratory, and has allowed the anxiety to sabotage any attempt to prepare intellectually. The result is that the student is defeated before ever starting the experiment.

Your chemistry laboratory experience will be enjoyable and rewarding if you learn to anticipate the challenges you will face and prepare to meet those challenges.

Some challenges to anticipate:

You will have difficulty using equipment or learning a particular technique.

Standard laboratory procedures such as lighting a burner, using a buret, and performing a vacuum filtration are described in this laboratory manual, and it is likely that your instructor or laboratory assistant will demonstrate these procedures for you in the laboratory when the need arises. Even with advance study and a demonstration, it is usually difficult to perform a new technique or use an unfamilar instrument correctly the very first time. *Anticipate* that you may feel clumsy or awkward at times in the laboratory just as you might on the first day of a new job. *With time and practice, you will become proficient and feel comfortable with the procedure.*

Some periods of time will be spent waiting.

During many experiments, you may have to wait to use a hood or balance, wait to have access to the chemicals needed for a reaction, and wait for water to boil or for a reaction to occur. You may start to feel bored and may even conclude that much of the lab time is being wasted by waiting *if* you have not *anticipated* these moments and *prepared* to use them. Making appropriate use of such waiting periods may be important to the successful completion of the experiment. Instead of standing by or chatting with others, you and your partner could be planning your stategy for the rest of the experiment, washing glassware needed for the next procedure, reviewing measurements and observations, or calculating a result.

Something will go wrong with your experiment.

In spite of your best efforts to prepare, sometimes things will go wrong in the laboratory: a beaker may shatter as you're washing it; a seemingly simple reaction may not give the expected result; your burner may not stay lit. Sometimes these incidents are the result of a mistake you have made: the reaction didn't work because the wrong chemicals were used (you didn't read the labels on the bottles), or the right chemicals were used but in the wrong proportions (you didn't follow directions carefully). Sometimes, however, the incident is the result of faulty laboratory equipment or reagents.

When something goes wrong, don't immediately assume that you have made a clumsy or careless error, but do not jump to the conclusion that the lab equipment or reagents are inferior. The goal in this situation is not to blame anything or anyone, but to try to determine what happened and why. You and your lab partner should examine the equipment carefully, try the reaction a second (or third) time, and discuss the possibilities with your instructor. If you find that you have made a mistake, learn from it, and try not make the same mistake a second time. If the error is with the equipment or chemicals provided to you, your instructor will replace these for you as quickly as possible.

If you've not anticipated and emotionally prepared for these moments, you may find that your anxiety level will rise or your temper will grow short. *Anticipate and prepare, and you will be able to patiently and calmly work with your partner or the instructor to solve any problem that arises.*

An effective way to prepare:

It is essential that you enter the laboratory fully prepared to perform each experiment. Anticipating the situations described above may help in your emotional preparation for laboratory, but each experiment will requre very careful, very thorough *intellectual* preparation. An excellent way to prepare intellectually is to create a "Pre-Laboratory Prep Sheet" for each experiment. The prep sheet is most useful if it is carefully organized so that is it informative and easy to follow.

Pre-Laboratory Prep Sheet

• *Objective and Important Procedural Steps*
Make a statement of what your goal is for the experiment: what do you hope to demonstrate or accomplish? List in a few sentences the key procedural steps you will take and the calculations you will perform to reach this goal.

• *Terminology*
List and define or explain the terms and procedures that are new, confusing, or difficult for you to remember, such as: formulas, structures, and names of chemicals; key terms and phrases; and special laboratory procedures.

• *Chemical Reactions and Mathematical Equations*
List balanced chemical equations for each reaction that you will perform. Make certain that you understand each equation and how it illustrates what you have learned in the lecture portion of the course: does the equation describe an acid-base reaction? an oxidation-reduction reaction?
List also all of the mathematical equations that you will need for calculating your results. Make certain that you know how to use the equations: what does each variable in the equation represent? what are the units of each variable? which values are measured? which values are given?

• *Laboratory Strategy*
List any special activities that you should do right away or a particular time during the experiment:

obtaining glassware and equipment from the stockroom or other location in the laboratory
washing and filling burets
setting up a boiling water bath or vacuum filtration apparatus
obtaining an "unknown" chemical from the instructor
weighing out a chemical for use in a reaction

• *Safety Tips and Special Considerations*
List any safety precautions that must be observed and special procedures that must be followed while performing the experiment. This list may include the following:

use of fume hoods
handling of chemicals that are flammable, corrosive, or have irritating vapors
proper disposal of chemical waste
careful check of labels on reagent bottles

• *Items to Bring to Lab*
List any items that you must bring with you when you come to the laboratory to perform the experiment. This list may include the following:

Pre-Laboratory Prep Sheet
preliminary calculations
laboratory manual
Periodic Table of the Elements
lab coat (or a change of clothes or shoes)
safety glasses
calculator
watch

Laboratory Safety Skills

Laboratory safety is a *skill;* it is, in fact, the most important skill that you must learn in the laboratory. Practicing safety skills means much more than learning and adhering to the list of safety regulations that are a part of every laboratory manual and introductory discussion. These lists usually address *personal* safety issues such as eye protection and appropriate clothing; you must remember, however, that *safety is social as well as personal.* The laboratory will be filled with students working on their experiments; you must learn to be conscious not only of your own actions, but also of those of the students around you. You must learn to concentrate on your own experiment while maintaining an awareness of the sights, sounds, and smells throughout the laboratory. *Before you act or react, give thought to the safety of others:*

• If you've broken a test tube or beaker, alert your neighbor to the broken glass, then quickly sweep up the glass. If your neighbor breaks some glassware, help him to alert others and offer assistance in the clean-up.

• If you've spilled a corrosive chemical, alert your neighbor to the spill and quickly clean it up *thoroughly* so that the next student to use your area will not be harmed by any residue. If your neighbor spills the chemical, help him to alert others, and offer assistance in the clean-up. Make certain that the clean-up is thorough.

• Before you light a burner, check for the presence of any flammable chemicals in your neighbor's work areas as well as your own. Do not leave your burner while it is lit, and pay close attention in case the burner should blow out while the gas remains on. When your neighbor prepares to light a burner, check about for the presence of flammable chemicals. Keep an eye on your neighbor's burner so that you will be aware if it blows out and the gas remains on.

• Know the locations of the laboratory exits, fire extinguishers, safety showers, eye-wash fountains, fire blankets, and emergency telephone; this knowledge may help you to save yourself *or your neighbor* from serious injury.

A list of laboratory safety regulations:

• *Students will work in the laboratory only during the scheduled period. Students are never permitted to work without supervision, and no unauthorized experiments are to be performed.*

• *Safety goggles, glasses, or other approved eye protection must be worn in the laboratory at all times regardless of what is being done.* Contact lenses are not acceptable eye protection. Students who do not follow this rule will be asked to leave the laboratory immediately.

• *Eating and drinking is prohibited in the laboratory;* the food can absorb toxic chemicals. *Smoking in or near the laboratory is prohibited;* flammable solvents can easily be ignited by a lighted cigarette.

• *Dress appropriately in the laboratory.* The feet must be adequately covered to prevent injury from dropped glassware or spilled chemicals; sandals and open-toed shoes are not acceptable. Clothing must be worn which covers the same parts of the body as a short-sleeve, full-length lab coat. Tanktops, shorts, and short skirts are not acceptable. Hair that is long enough to interfere with an experiment should be tied back.

• *Know the location of all safety equipment in the laboratory and understand how the equipment is operated:*

fire extinguishers
fire blankets
safety showers
eye wash fountains.

• *Know the location of all laboratory exits.*

• *Know the location of the emergency telephone and the number to call security or emergency personnel.*

• *Never smell any chemical unless instructed to do so,* and then waft the vapors toward your nose with your hand. *Use dangerous or irritating chemicals in the fume hood.*

• *Never taste any chemical, and never pipet by mouth.* Some chemicals are very corrosive or poisonous even in very small quantities.

• *If chemicals come into contact with the hands or body, the area should be rinsed immediately and liberally with water and the intructor notified. Chemical spills in the work area should be cleaned up immediately as directed by the instructor.*

• *Never store a chemical in your laboratory drawer unless you are directed to do so.*

• *Never remove anything (chemicals, glassware, etc.) from the laboratory.* It is illegal to do so.

• *Never mix chemicals arbitrarily or carelessly.* Read labels carefully to make certain that you have the correct chemicals and in the correct concentrations.

• *Never put anything back into a reagent bottle;* take what is needed for a given procedure and discard any excess in the appropriate waste container.

• *Do not dispose of chemicals in laboratory sinks or trash cans.* Only soap and water are to be put into the sinks. Specially marked waste containers will be provided for each chemical, and students should carefully follow the instructor's directions for waste disposal.

• *Open flames are prohibited in the laboratory unless specifically called for in the experiment. A lighted Bunsen burner is never to be left unattended.*

• Even when the best of precautions are taken, accidents may occur, and should not be a cause for embarassment. *All accidents and injuries, even minor ones, must be reported to the instructor immediately.*

• *At the completion of the laboratory period, clean your working area, return special equipment to its proper place, make certain that water and gas valves are shut off, thoroughly wash your hands and safety glasses.*

Skills in Recording Data and Working Calculations

The recording of numerical data and experimental observations is another laboratory skill that requires thought and practice. Students sometimes forget to write down a measurement or an observation, and they often feel that they can simply *remember* what they have observed and record it in their notebooks at a later time. So many things occur during the laboratory period that you not be able to remember everything. You must concentrate on recording (legibly, and in ink) all measurements and observations, *as they are made,* on the laboratory report sheet or in a laboratory notebook. Keep in mind that *you can never write down too much;* most of the questions that arise as students prepare their reports are the result of improperly recorded data and insufficient observation. (Realize that if you have done a thorough intellectual preparation for the experiment, you will *know* what to look for and what to record.)

When making observations of a substance, describe everything that you see (and perhaps smell): Is the substance a solid, liquid, or gas? What color is it? Is it opaque, translucent, or transparent? If you are instructed to smell the substance, what is its odor?

When making observations of a reaction, write down what you are doing, and then what you see, hear, smell, and feel: Is there a color change? Is a gas evolved? Does a solid form? Is an odor emitted? Is there any sound? Is any heat evolved? How long does it take for the reaction to occur?

Be careful that your observation is indeed an observation and not an inference. Write down only what you've actually observed, and not what you *think* you've observed. Example: A metal can be described as a *silver colored metal,* but should not be described as *silver metal.*

Numerical data must be recorded *with units, and to the appropriate number of significant figures.*

When you are asked to show a calculation of a result, first write the mathematical equation that is to be used, then substitute in the appropriate values with their units, and finally show the result with the correct units and significant figures.

Other Laboratory Skills

Measurement of Volume: Use of Graduated Cylinders and Burets

Laboratory glassware is calibrated in terms of the volume contained or delivered: each piece is labeled either "TC" (to contain) or "TD" (to deliver). These calibrations differ by a small but significant amount. Because the volume of a given mass of liquid varies with temperature, most glassware is calibrated for use with liquids at 20.0 °C, an average room temperature.

It is important that glassware be kept clean. Immediately after use, the container should be emptied, rinsed several times with tap water, washed with dilute detergent solution, rinsed with tap water again, and finally rinsed several times with distilled water. After rinsing, a film of water should adhere uniformly to the inside surface of the container. If the wetting appears uneven or spotty, the container needs additional cleaning. It is best to let a container dry by standing at room temperature.

Graduated Cylinders

Graduated cylinders vary in capacity from 5 mL to about 2000 mL. Most are calibrated "to deliver" and have a pouring spout. The volume scale increases from the bottom to the top.

The surface of a liquid in a cylinder will exhibit a definite curvature called a meniscus. For most liquids, this is a concave curvature, and the bottom of the meniscus is used to read the volume. Some liquids (mercury, for example) have a convex surface; the top of this meniscus is used for volume readings.

The eye must be level with the surface of the liquid to observe the correct volume. If the eye is above the meniscus, too large a volume is observed; if the eye is below the meniscus, the volume reading will be too small. This apparent variation in the volume as observed from different locations is called "parallax."

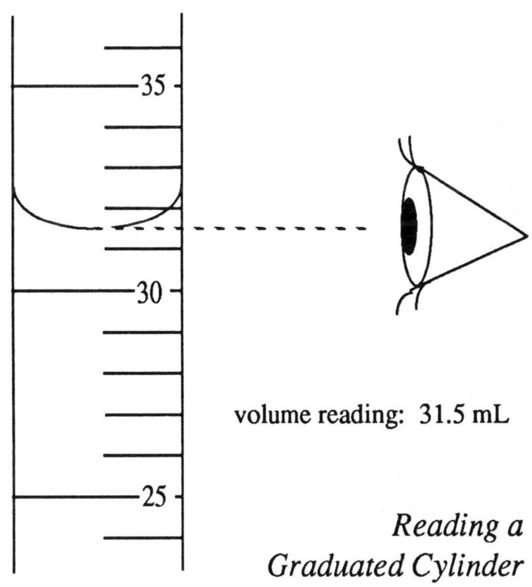

Reading a Graduated Cylinder

Burets

A buret is a long, narrow, calibrated tube with a valve ("stopcock") at one end to control the flow of liquid. Common burets sizes are 10, 25, 50, and 100 mL.

Burets are used to make precise measurements of the volume of a liquid that is delivered into another container. If this volume measurement is to be accurate, the buret must be kept extrememly clean. The inner walls can be scrubbed with a buret brush that has been dipped in a dilute detergent solution. The buret should be thoroughly rinsed, first with tap water and then with distilled water.

When the buret is clean and ready for use, it should be rinsed several times with small portions of the solution to be delivered, then filled to above the zero calibration at the the top of the buret. A funnel can be used for this, but the solution should be poured slowly to avoid creating air bubbles and overfilling the buret. The stopcock is opened to carefully lower the meniscus to the zero calibration or below; any air bubbles in the tip of the buret should be flushed out by this action. A drop of liquid hanging from the tip of the buret should be dislodged; the initial volume of the solution in the buret must be read and recorded.

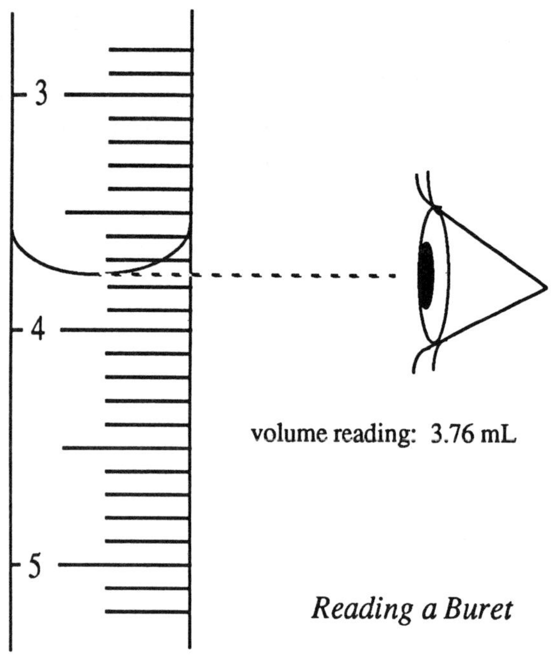

Reading a Buret

Care should be taken in reading the liquid level in a buret. A 50 mL buret is marked every 0.1 mL, but fifths of each division can be estimated with reproducibility; each volume should be recorded to the nearest 0.01 mL.

Burets should be handled carefully to avoid dislodging the stopcock. The best method for withdrawing liquid from a buret is as follows:

- the stopcock valve handle is at the right of the buret;
- the left hand reaches behind the buret (thumb in front) to operate the stopcock;
- the right hand holds and swirls the receiver flask.

After the desired volume is withdrawn, the receiver flask should be touched to the buret tip to dislodge the final drop. After a pause (to allow the film of liquid on the inner walls to drain to the meniscus), the final volume should be read and recorded.

After use, the buret should be drained and rinsed thoroughly.

Separation of Precipitates: Decantation, Gravity Filtration, and Vacuum Filtration

Decantation

Very dense precipitates will settle to the bottom of a container; the clear solvent above the precipitate is then simply decanted (poured off), leaving the precipitate in the original container. The precipitate may be washed by adding fresh solvent, allowing the solid to settle, and decanting again.

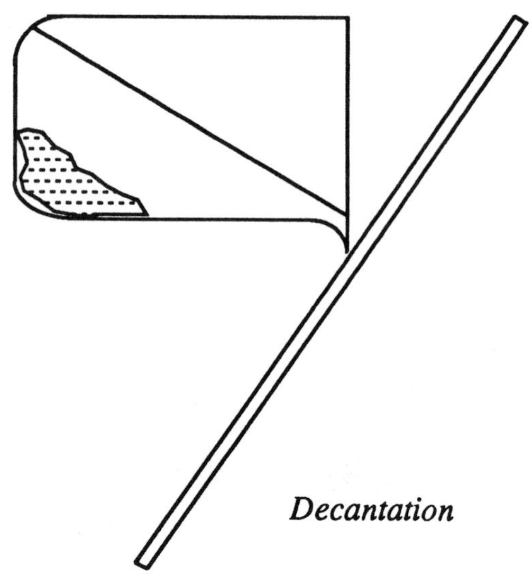

Decantation

Gravity Filtration

Gravity filtration is commonly used to separate insoluble impurities from an important solution.

The precipitate and solvent are poured into a funnel fitted with a sheet of porous paper through which the solvent can pass but the precipitate cannot. The filter paper is folded in half, in half again (so that the edges do not match), and a corner of the smaller portion is torn off.

The larger portion is opened, and the resulting cone is fitted into a funnel and moistened with solvent to form a seal. It is essential that the paper fit properly so that the funnel will drain well.

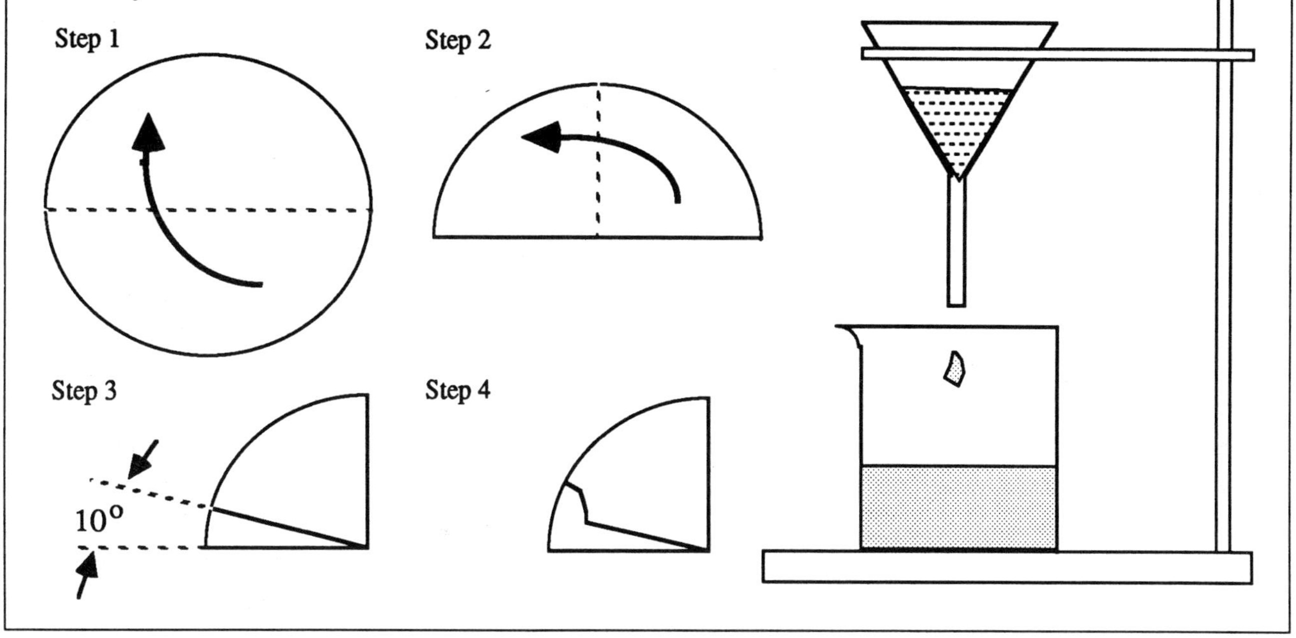

Gravity Filtration

Vacuum Filtration

Vacuum filtration is frequently used to separate an important precipitate from a discardable solvent.

The vacuum is produced by a water aspirator at the lab bench (the faucet with a short horizontal tube). A rapid flow of water inside the main pipe draws air through the horizontal tube, generating a partial vacuum.

The filter flask should be clamped to a ring stand before the Buchner funnel is put in place. The funnel makes the apparatus top heavy, and if not clamped, it may fall over.

A filter paper slightly smaller than the perforated plate of the funnel is put into place and moistened with solvent. It is important that the wet paper form a seal against the funnel when the vacuum is turned on; otherwise, it may float when the solid-solvent mixture is poured into the funnel.

The mixture should be poured slowly into the center of the funnel. When the filtration is

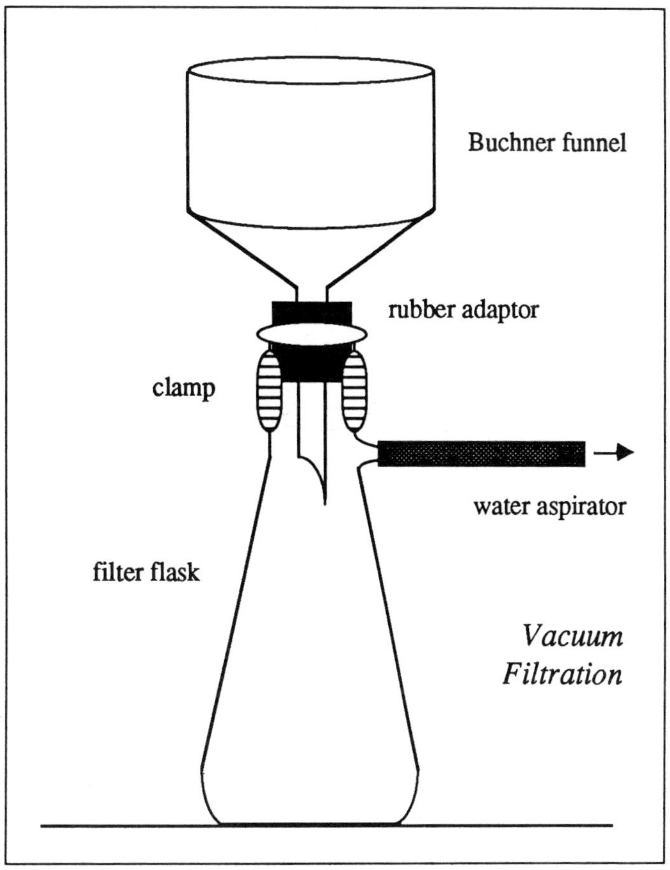

Vacuum Filtration

times with fresh solvent. The vacuum should be shut off (this should be done by pulling the tubing from the filter flask, and *then* turning off the water), and the filter cake broken up with a spatula. A small volume of solvent is added, and the mixture stirred in the funnel. This must be done carefully so that the filter paper is not torn or disturbed. The vacuum is turned on again, and the solvent removed. After several washes, the vacuum can be used to pull air through the filter cake to dry the solid.

Operation of a Bunsen Burner

The burner is most easily lit if the gas flow rate is intermediate and the amount of air in the gas-air mixture is low. If the burner has a threaded base, the air intake can be reduced by tightening the rotatable barrel all the way to close off the air ports, then loosening the barrel just one to two turns. If the burner has a rotatable sleeve over the air ports at the base, the sleeve should be rotated so that the ports are nearly covered.

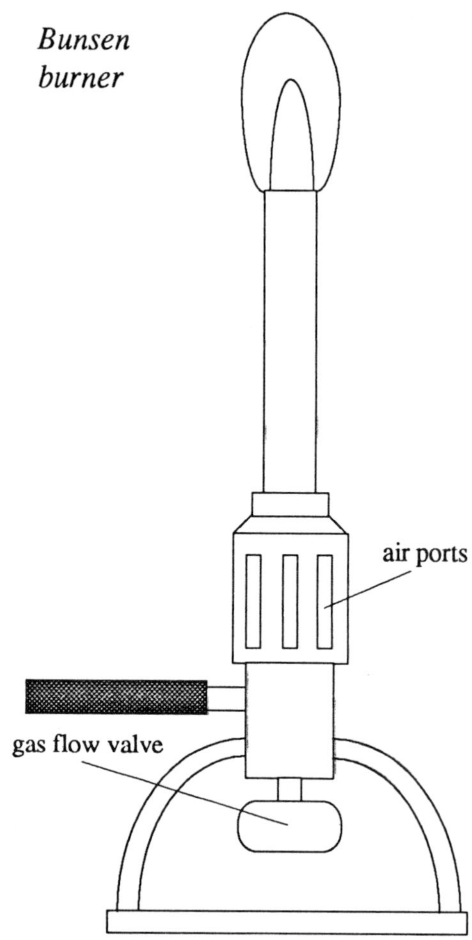

Bunsen burner

The rubber hose on the burner is connected to the bench-top gas supply valve. The gas flow valve at the base of the burner should be turned to the fully-closed position, and the main gas valve then turned to the fully-opened position. The gas flow is then controlled from the valve at the base of the burner.

When lighting a burner with a match, *the match should be lighted first, before the burner gas valve is opened one to two turns*. If the burner does not light easily, the gas should be quickly shut off, and the instructor consulted.

The *height* of the flame can be adjusted by changing the gas flow; the *temperature* of the flame is adjusted by altering the air flow. Too little air will produce a luminous (yellow), low temperature flame. Increased air flow produces a high temperature, non-luminous flame with an inner blue cone. The hottest part of this flame is at the tip of the blue cone. A noisy flame signals that the air flow is too high.

Before using a burner, make certain that no flammable solvents are being used nearby. A student with long hair should tie the hair back, and loose sleeves should be pushed back on the arms.

Experiment 1

Structure, Isomerism, and Nomenclature of Organic Compounds

What you will accomplish in this experiment

This exercise will give you an opportunity to experience the three-dimensional nature of molecules and to visualize how structural formulas are used to represent three-dimensional molecules. You will become better acquainted with structural and stereoisomerism among organic compounds. Additionally, the exercise will demonstrate how organic nomenclature specifies the composition and structure of molecules, and even the spatial arrangement of atoms within a molecule.

Concepts to know to be prepared

Before coming to laboratory, you should consult your lecture notes and carefully read the appropriate sections of your textbook to become familar with the following concepts:

- structures of alkanes, alkenes, and cycloalkanes
- complete and condensed structural formulas
- general rules of organic nomenclature
- structural isomerism
- *cis* and *trans* stereoisomerism

Procedure that you will follow

Each pair of students will be supplied with a set of molecular models. Each set should contain the following:

30 tetrahedral carbon atoms (black)
6 pairs of double-bonded carbon atoms (black)
5 oxygen atoms (red)
2 nitrogen atoms (blue).

Construct a model for each of the molecules represented on the report sheet.

Report Sheet: Experiment 1 Structure, Isomerism, and Nomenclature of Organic Compounds

Name ____________________________________ *Date* ____________ *Section* _________

Partner's Name ___________________________ *Instructor's Initials* ___________________

A. Alkanes

1. Construct a model of ethane, C_2H_6. Note the tetrahedral geometry of the electron pairs about the carbon atoms, and that the two methyl groups can be rotated about the carbon-carbon single bond without breaking the bond.
Write the complete and condensed structural formulas for ethane.

2. Construct a model of propane, C_3H_8. Note that because of the tetrahedral geometry of the carbon atoms, the carbon skeleton is not linear, but has a zig-zag appearance.
Write the complete and condensed structural formulas for propane.

B. Structural Isomerism of Alkanes

1. Construct models for the three molecules that have the molecular formula C_5H_{12}. These molecules are *structural isomers.*
Write condensed structural formulas and give IUPAC names for these molecules.

C. Stereoisomerism of Cycloalkanes

1. Construct a model of cyclopentane, C_5H_{10}. Note that the ring is planar. Now attach a methyl group to each of two adjacent carbons to create a molecule of 1,2-dimethylcyclopentane, C_7H_{14}. Note that the methyl groups can be attached in two different ways:
   1) both methyl groups on the same side of the cycloalkane ring, and
   2) methyl groups on opposite sides of the cycloalkane ring.

These are the *cis* and *trans stereoisomers,* respectively.
Write condensed structural formulas for these molecules, and label the structures as *cis* or *trans* as appropriate.

D. Alkenes

1. Examine the model of ethene, C_2H_4. This is the piece in the model kit that contains the carbon-carbon double bond; it is understood that hydrogens occupy positions at the ends of the single bonds. Note the trigonal planar geometry of the electron pairs about the carbon atoms, and that the carbon-carbon double bond cannot be rotated.
Write the complete structural formula for ethene.

2. Construct a model of propene, C_3H_6. Note the trigonal planar geometry of the two carbon atoms participating in the double bond, and the tetrahedral geometry of the third carbon atom.
Write the complete structural formula for propene.

E. Structural and Stereoisomerism of Alkenes

1. Start with a model of ethene, C_2H_4, and attach a methyl group to each of the two carbons to create a molecule of 2-butene, C_4H_8. Note that the methyl groups can be attached in two different ways:

   1) both methyl groups on the same side of the double bond, and
   2) methyl groups on opposite sides of the double bond.

These are the *cis* and *trans stereoisomers,* respectively.
Write structural formulas for these molecules, and label the structures as *cis* or *trans* as appropriate.

2. Two additional molecules have the molecular formula C_4H_8. These molecules are structural isomers of one another and of the stereoisomers in part E.1. Construct models of these molecules, then write their condensed structural formulas and give the IUPAC name for each molecule.

F. Additional Exercises in Structure and Nomenclature

1. Write a condensed structural formula for each of the following compounds. You will find it helpful to build a model of each molecule.

a. 2,4-dimethylhexane

b. 2,3,5-trimethyl-4-propyl-1-heptene

c. 3-ethyl-4-isopropylheptane

2. Give the IUPAC name for each of the following compounds. Be certain that you have identified the longest carbon chain in each molecule. You will find it helpful to construct a model of each molecule.

a. 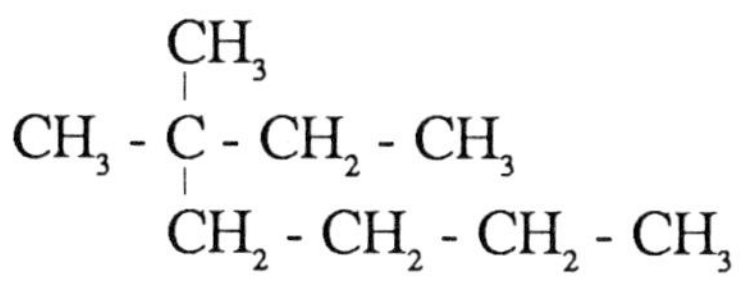

b. 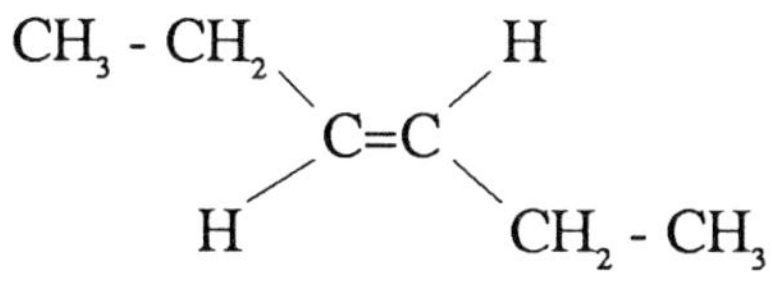

c. 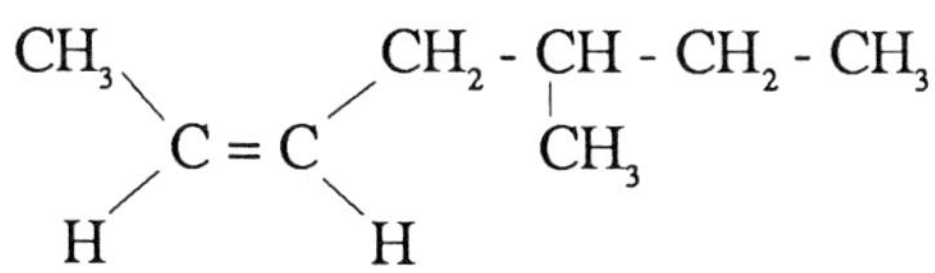

d. 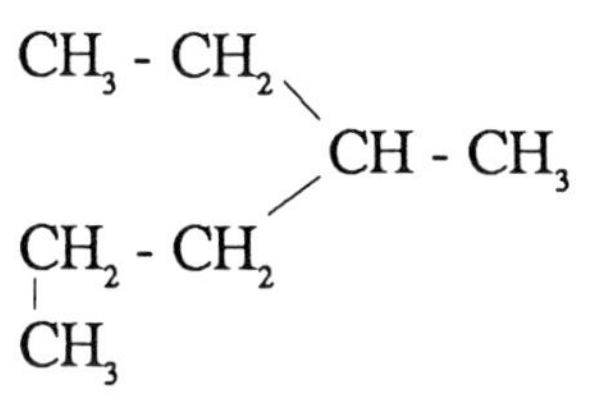

e. 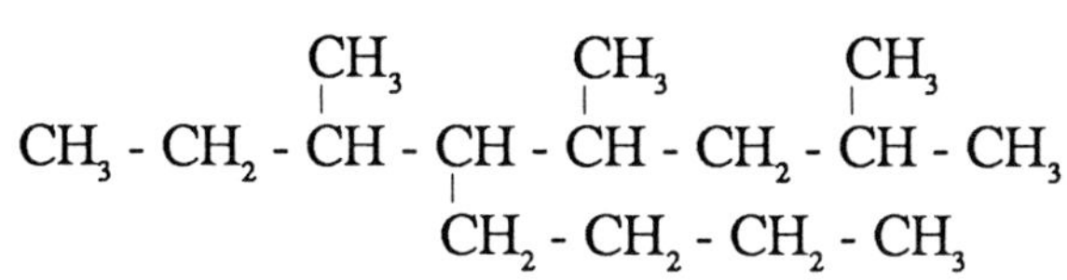

f. 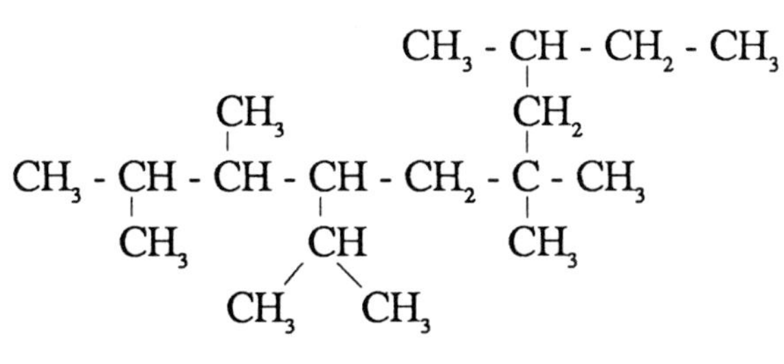

Experiment 2

Organic Functional Group Analysis

What you will accomplish in this experiment

This experiment will introduce you to some of the more common functional groups of organic chemistry. The functional group is that portion of the molecule that undergoes a structural change during a chemical reaction. The functional groups that will be studied in this experiment are carboxylic acid, amine, aldehyde, ketone, alcohol, and alkene.

You will learn chemical tests that will allow you to distinguish one functional group from another. You will use the chemical tests to identify the functionality of an unknown organic compound. In addition, you will use a water solubility test to determine whether your organic compound is of high or low formula weight.

Concepts to know to be prepared

The chemical tests you will perform make up a sequence of experiments designed to determine the absence of or suggest the presence of particular functional groups. The complete sequence is shown in the flow diagram on page 24. This diagram can serve you in several ways:

- It is a summary of the procedure that you are to follow in classifying your unknown as one of the functional group types.

- It can order your thoughts as you read the discussion of each test, and help you to understand the significance of that test.

- It can enhance your appreciation for and enjoyment of this experiment. Your role is that of chemist and detective: you will employ this cleverly devised scheme to sleuth out the identity of your unknown's functionality.

Discussion of Chemical Tests

1. Water Solubility: Diagnosis of a Low Formula Weight Compound

Most organic compounds are slightly soluble or insoluble in water. Water solubility is uncommon, and if observed, is an indication of a molecule of low formula weight. A litmus paper test of the solution's pH will provide additional information about the water-soluble molecule.

A. Acidic Response to Litmus: Diagnosis of a Low Formula Weight Organic Acid

Carboxylic acids with fewer than five carbon atoms are generally water soluble and form solutions that give an acidic response to litmus (acidic response: blue litmus paper turns red).

$$R-\overset{O}{\overset{\|}{C}}-O-H + H_2O \rightleftharpoons R-\overset{O}{\overset{\|}{C}}-O^- + H_3O^+$$

B. Basic Response to Litmus: Diagnosis of a Low Formula Weight Organic Base

Amines are the organic derivatives of ammonia, and like ammonia, they are weak bases. Amines that have fewer than five carbon atoms are generally water soluble and form solutions that give a basic response to litmus (basic response: red litmus paper turns blue).

primary amine: $RNH_2 + H_2O \rightleftharpoons RNH_3^+ + OH^-$

secondary amine: $R_2NH + H_2O \rightleftharpoons R_2NH_2^+ + OH^-$

tertiary amine: $R_3N + H_2O \rightleftharpoons R_3NH^+ + OH^-$

C. Neutral Response to Litmus: Diagnosis of Low Formula Weight Neutral Compound

Neutral compounds (ketones, aldehydes, alcohols, alkenes) with fewer than five carbons are generally water soluble, and the resulting solutions are neither acidic nor basic. The solutions give a neutral response to litmus (neutral response: litmus paper does not change colors).

2. Solubility in 5% NaOH: Diagnosis of a High Formula Weight Organic Acid

A carboxylic acid with five or more carbons, though insoluble in water, will be soluble in 5% NaOH solution because of the formation of the sodium salt of the acid.

$$R-\overset{O}{\overset{\|}{C}}-O-H + NaOH \longrightarrow R-\overset{O}{\overset{\|}{C}}-O^-Na^+ + H_2O$$

3. Solubility in 5% HCl: Diagnosis of a High Formula Weight Organic Base

An amine with five or more carbon atoms, though insoluble in water, will be soluble in 5% HCl solution because of the formation of an ammonium chloride salt.

primary amine: $RNH_2 + HCl \longrightarrow RNH_3^+Cl^-$

secondary amine: $R_2NH + HCl \longrightarrow R_2NH_2^+Cl^-$

tertiary amine: $R_3N + HCl \longrightarrow R_3NH^+Cl^-$

If the compound is not soluble in either dilute acid or dilute base, then it is a neutral compound (ketone, aldehyde, alcohol, alkene) of high formula weight.

4. Reaction with 2,4-Dinitrophenylhydrazine (2,4-DNP): Diagnosis of Aldehyde or Ketone

Aldehydes and ketones are detected by the formation of an orange or red-orange 2,4-dinitrophenylhydrazone precipitate.

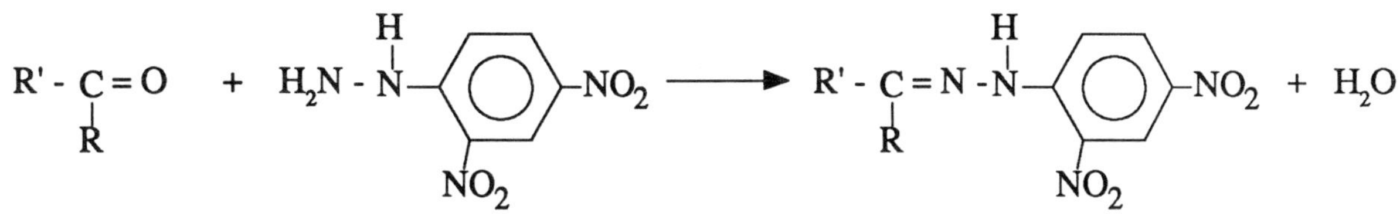

R' = alkyl or H 2,4-DNP orange 2,4-DNP precipitate

5. Tollen's Test: Diagnosis of Aldehyde

Aldehydes can be distinguished from ketones by the Tollen's Test, because aldehydes can be easily oxidized to acids, but ketones cannot. The Tollen's reagent contains a silver-ammonia complex ion in a basic solution. Silver(I) ion is reduced to silver metal as the aldehyde is oxidized to acid.

$$R-\overset{O}{\overset{\|}{C}}-H + 2\,Ag(NH_3)_2^+ + OH^- \longrightarrow R-\overset{O}{\overset{\|}{C}}-O^- + 2\,Ag + H_2O + 3\,NH_3$$

An aldehyde is indicated by the appearance of a silver mirror on the wall of the test tube, or a precipitate of silver metal.

6. Reaction with Chromic Acid: Diagnosis of Alcohol

Primary and secondary alcohols are oxidized by chromic acid to carboxylic acid and ketone, respectively.

primary alcohol: $$R-CH_2-O-H \xrightarrow{H_2CrO_4} R-\overset{O}{\overset{\|}{C}}-H \xrightarrow{H_2CrO_4} R-\overset{O}{\overset{\|}{C}}-O-H$$

secondary alcohol: $$R'-\underset{R}{\underset{|}{\overset{H}{\overset{|}{C}}}}-O-H \xrightarrow{H_2CrO_4} R'-\overset{O}{\overset{\|}{C}}-R$$

tertiary alcohol: $$R_3C-O-H \xrightarrow{H_2CrO_4} \text{no visible reaction}$$

An alcohol is indicated by disappearance of the orange color of the chromic acid reagent and the formation of a green or blue-green solution.

7. Decolorization of Bromine: Diagnosis of Alkene

The halogen bromine will rapidly react with alkenes at room temperature to produce the dibromoalkane.

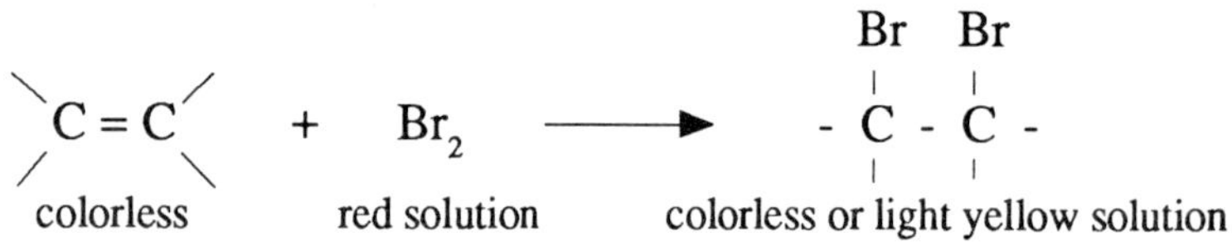

Bromine solutions are reddish-brown; alkenes and bromoalkanes are colorless. Rapid decolorization of bromine is a positive indication of an alkene.

Procedure that you will follow

The following procedures are to be carried out first on samples of known compounds so that positive responses for each test can be observed. When you feel confident that you can recognize a positive test *for each procedure,* obtain an unknown from your instructor. Substitute your unknown for the known compound in each test. Use the flow diagram on page 24 to follow your progress in identifying the functionality and formula weight (high or low) of your unknown.

Known compounds to be tested for solubility properties: ethanoic acid, propanone, cyclohexene, diethylamine, butanal, 1-pentanol.

1. Water Solubility: Diagnosis of a Low Formula Weight Compound

Add twenty drops of water to ten drops of the known compound. Agitate the mixture thoroughly. A liquid that is insoluble in water will form a second phase.
Visualization of two liquid phases in a few drops of mixture can be difficult; holding the test tube nearly horizontal should make the liquid-liquid interface readily observable.
If the compound is soluble in water, it is a low formula weight compound, and the solution should be tested with litmus paper to determine whether the compound is a carboxylic acid, an amine, or a neutral compound (ketone, aldehyde, alcohol, or alkene).
If the compound is insoluble in water, it is a high formula weight compound. The compound's solubility properties in 5% NaOH and 5% HCl will yield information about its functionality.
Record your observations and draw conclusions about each compound tested.

2. Solubility in 5% NaOH: Diagnosis of a High Formula Weight Organic Acid

Add twenty drops of 5% NaOH dropwise, with shaking, to ten drops of the known compound.
A high formula weight carboxylic acid will be soluble in 5% NaOH because of the formation of the sodium salt of the acid. High formula weight amines and neutral compound will not be soluble in 5% NaOH; these functionalities will be distinguished by their differing solubilities in 5% HCl.
Record your observations and draw conclusions about each compound tested.

3. Solubility in 5% HCl: Diagnosis of a High Formula Weight Organic Base

Add twenty drops of 5% HCl dropwise, with shaking, to ten drops of the known compound.

High formula weight amines will be soluble in 5% HCl because of the formation of the ammonium chloride salt. High formula weight neutral compounds (ketones, aldehydes, alcohols, alkenes) will not be soluble in 5% HCl.

Record your observations and draw conclusions about each compound tested.

4. Reaction with 2,4-Dinitrophenylhydrazine (2,4-DNP): Diagnosis of Aldehyde or Ketone

Known compounds to be tested: propanone, butanal

Place twenty drops of the 2,4-DNP solution in a small test tube. Add two drops of the known compound and swirl the mixture.

Aldehydes and ketones will react with the reagent and produce an orange or red-orange precipitate.

Record your observations for each compound tested.

5. Tollen's Test: Diagnosis of Aldehyde

Known compounds to be tested: propanone, butanal

Place forty drops of 5% aqueous $AgNO_3$ in a medium-size, *very clean* test tube. (Obtain a new test tube from the stockroom for this test.) Add one drop of 10% NaOH solution; a powdery brown precipitate of silver oxide should form. Add 5% NH_3 dropwise, with agitation, until the muddiness of the mixture clears. (Larger pieces of the brown solid will likely remain near the bottom of the test tube.) *Do not use excess ammonia.* Add to this mixture *one drop* of the known compound and shake thoroughly. Warm the solution in a hot water bath (50-60 °C, *no hotter*) for several minutes.

An aldehyde is indicated by the formation of a silver mirror on the wall of the test tube.

Record your observations for each compound tested.

Be sure to follow these directions carefully. Excess ammonia, too much unknown, overheating, or a dirty test tube will all give poor results. Upon completion of the test, wash away the reaction mixture with plenty of water. It may deposit explosive products upon standing or drying out.

6. Reaction with Chromic Acid: Diagnosis of Alcohol

Known compound to be tested: 1-pentanol

Place twenty drops of the known compound in a test tube. Add 1 drop of the chromic acid reagent and swirl the mixture.

An alcohol is indicated by the disappearance of the orange color of the chromic acid reagent and the formation of a green or blue-green solution.

Record your results for each compound tested.

7. Decolorization of Bromine: Diagnosis of Alkene

Known compound to be tested: cyclohexene

Place two drops of the known compound in a *dry* test tube. Add dropwise a 1% solution of Br_2 in carbon tetrachloride (CCl_4).

An alkene will rapidly decolorize at least ten drops of the bromine solution.

Record your results for each compound tested.

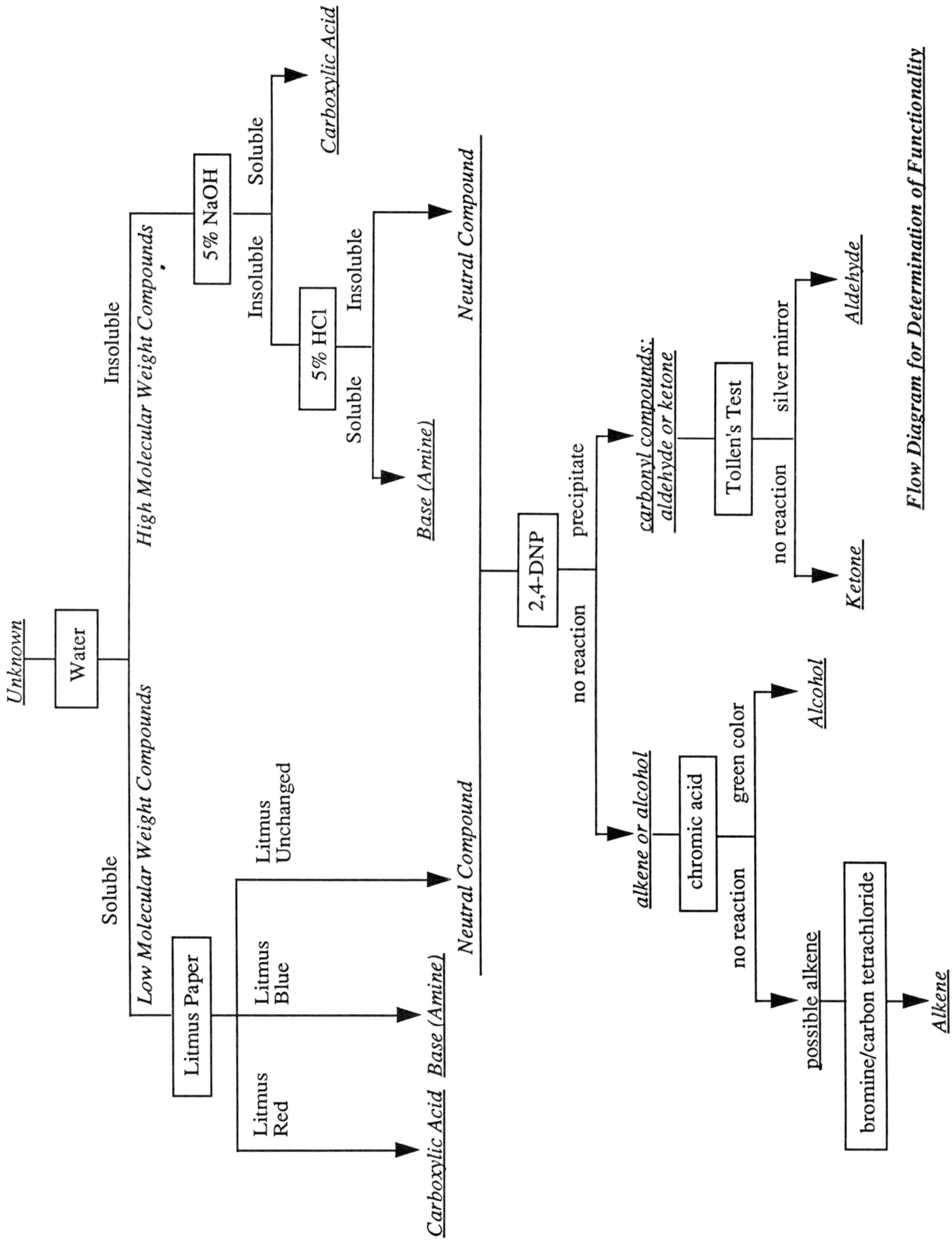

Flow Diagram for Determination of Functionality

Report Sheet: Experiment 2
Organic Functional Group Analysis

Name ______________________ *Date* __________ *Section* __________

Partner's Name ______________________ *Instructor's Initials* ______________

KNOWN Compound Functional Group Tests

Tests 1, 2, 3. Solubility

Known Compounds	Observations			Conclusions
	Water/Litmus	5% NaOH	5% HCl	
cyclohexene				
diethylamine				
propanone				
1-pentanol				
butanal				
ethanoic acid				

Test 4. Reaction with 2,4-Dinitrophenylhydrazine

Known compound tested: propanone

Observations:

Chemical equation:

Test 4, continued. Reaction with 2,4-Dinitrophenylhydrazine

Known compound tested: butanal
Observations:

Chemical equation:

Test 5. Tollen's Test

Known compound tested: propanone
Observations:

Chemical equation:

Known compound tested: butanal
Observations:

Chemical equation:

Test 6. Reaction with Chromic Acid

Known compound tested: 1-pentanol
Observations:

Chemical equation:

Test 7. Decolorization of Bromine

Known compound tested: cyclohexene
Observations:

Chemical equation:

UNKNOWN Compound Functional Group Tests

Unknown Number ____________________

Chemical Test	Observations	Conclusions
Water Solubility/Litmus		
5% NaOH Solubility		
5% HCl Solubility		
Reaction with 2,4-Dinitrophenylhydrazine		
Tollen's Test		
Reaction with Chromic Acid		
Decolorization of Bromine		

Functionality of Unknown ____________________________

Formula Weight of Unknown (high or low) _______________

Experiment 3

Synthesis of Aspirin

What you will accomplish in this experiment

You will:

- synthesize acetylsalicylic acid (aspirin) by carrying out a simple organic reaction,
- separate your product from the reaction mixture by vacuum filtration,
- purify your product by recrystallization,
- perform a chemical test to identify the change in functional group from reactant to product, and
- determine the success of your synthesis by calculating the percentage yield of your product.

Concepts to know to be prepared

Aspirin (acetylsalicylic acid), first synthesized in 1899, is just one of the many derivatives of *salicylic acid* that have important medical applications. Aspirin serves as an analgesic (pain killer), an antipyretic fever depressant), and an antiseptic (germicide).

Salicylic acid is both an acid and a phenol, because it has a carboxyl and a hydroxyl group attached to a benzene ring. It is a colorless, crystalline solid.

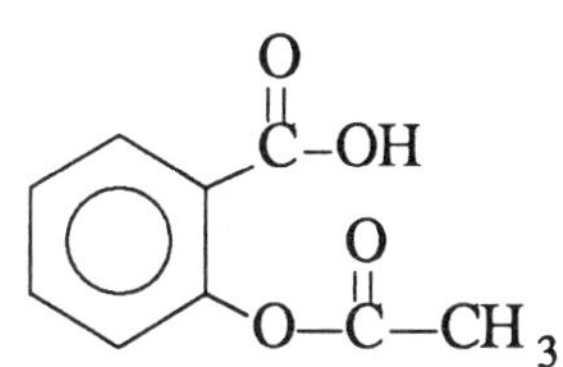

aspirin (acetylsalicylic acid)

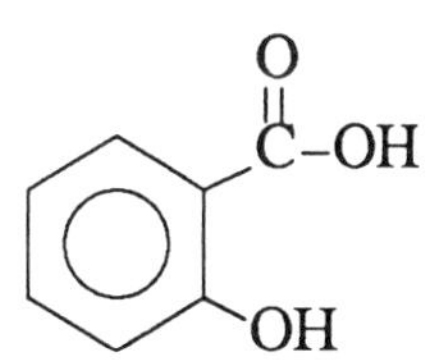

salicylic acid

Derivatives of Salicylic Acid

1. Salts

When salicylic acid is treated with sodium hydroxide, the salt sodium salicylate is the product of the acid-base reaction. This salt is used as a food preservative and in the treatment of arthritis.

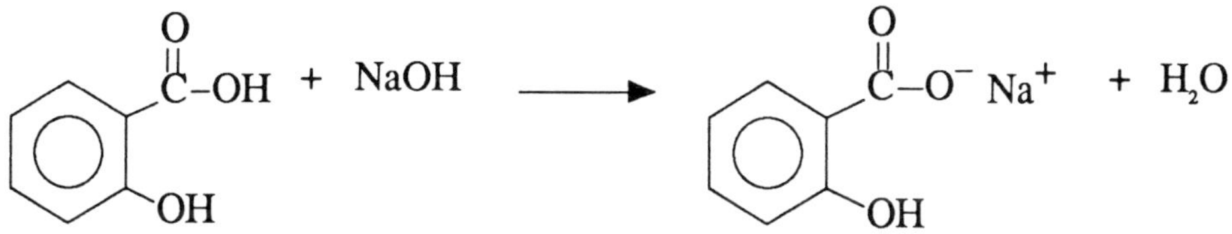

salicylic acid → sodium salicylate

2. Esters

When a carboxylic acid reacts with an alcohol or phenol under acidic conditions, an ester and water are produced from the condensation reaction:

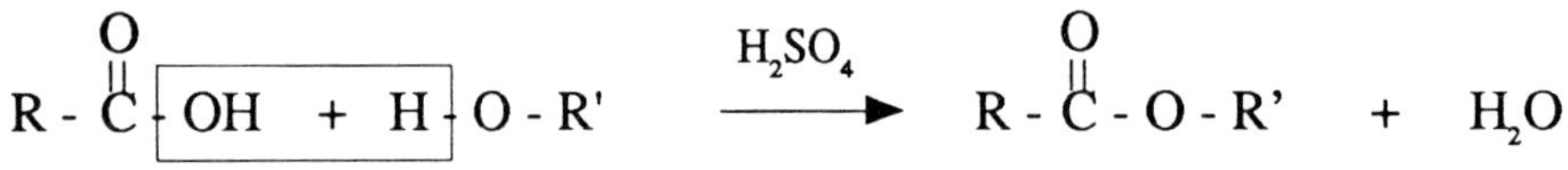

carboxylic acid | alcohol | ester | water

The methyl ester of salicylic acid, known as "oil of wintergreen," is used in liniments, perfumes, and flavorings. Methyl salicylate (oil of wintergreen) is prepared by the following condensation reaction:

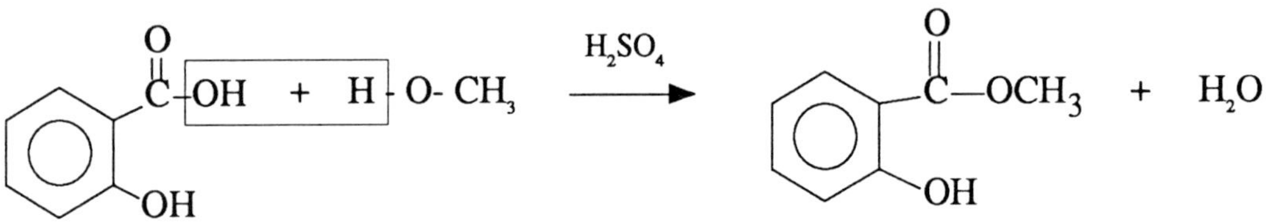

salicylic acid | methanol | methyl salicylate, $C_8H_8O_3$ | water

The phenol ester of salicylic acid, "salol," is used as an intestinal antiseptic. It passes through the stomach unchanged and is slowly hydrolyzed to phenol and salicylic acid in the small intestine. Salol is used to coat pills that are to reach the intestines unchanged.

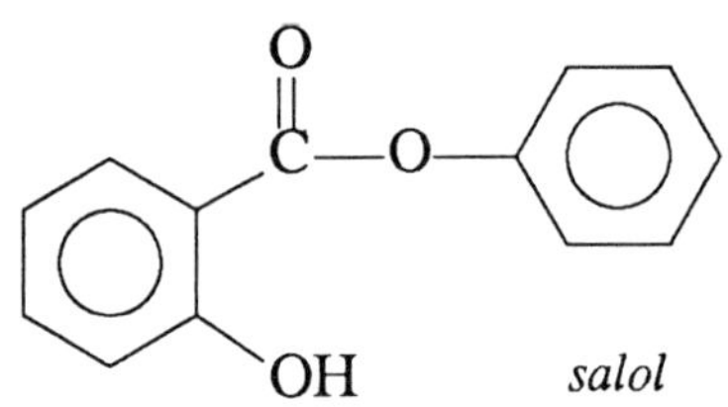

salol

Aspirin can be produced by reacting salicylic acid with glacial acetic acid; however, this condensation reaction occurs very slowly.

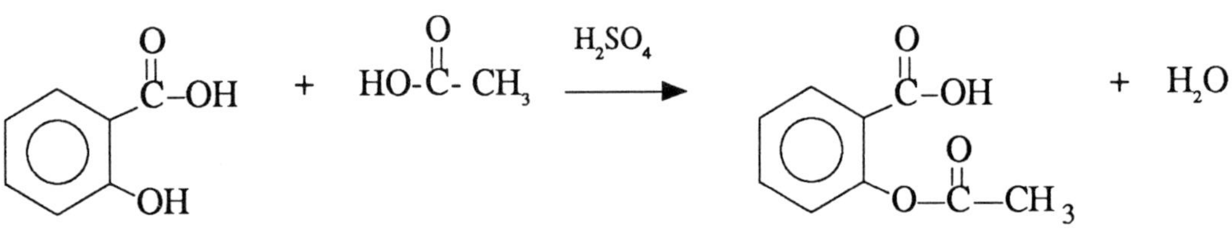

salicylic acid | acetic acid | acetylsalicylic acid (aspirin) | water

In this experiment, aspirin will be produced by reacting salicylic acid with acetic anhydride:

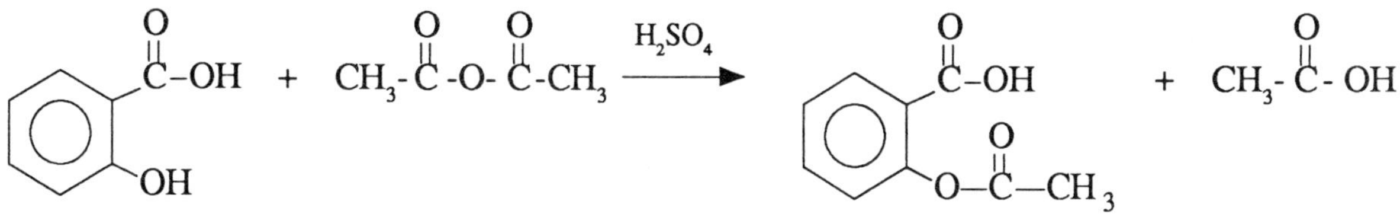

salicylic acid	acetic anhydride	aspirin acetylsalicylic acid	acetic acid
$C_7H_6O_3$	$C_4H_6O_3$	$C_9H_8O_4$	$C_2H_4O_2$
FW = 138.00 g/mol	FW = 102.10 g/mol	FW = 180.17 g/mol	FW = 60.06 g/mol

The acetic anhydride molecule is split in this reaction. One part, the acetyl group, $CH_3-\overset{O}{\overset{\|}{C}}-$, is substituted *for* the phenolic hydrogen to form acetylsalicylic acid (aspirin).

The other half, the acetate group, $CH_3-\overset{O}{\overset{\|}{C}}-O^-$, combines *with* the phenolic hydrogen ion to form acetic acid.

Recrystallization

A chemical reaction seldom occurs in such a manner that the desired product is obtained in an absolutely pure state. The synthetic reaction in this experiment will be no exception. Consequently, a purification procedure must be carried out to separate the aspirin from such impurities as unused starting materials and the acetic acid by-product.

Recrystallization is an extremely useful technique for purifying crystalline solids. The impure product is dissolved in a minimum volume of hot solvent. The solvent is carefully selected so that the product is much less soluble in cold solvent than hot, so that the product will recrystallize when the solvent is cooled, leaving the impurities behind in the solvent. The pure product is then isolated by vacuum filtration.

Iron (III) Chloride Test for Phenols

A convenient test can be carried out to identify the change in functional group from reactant to product. A comparison of the structural formulas of aspirin and salicylic acid shows that the salicylic acid has a phenolic group while aspirin has none. A specific test for a phenolic group will illustrate this change in functionality.

If a drop of iron (III) chloride ($FeCl_3$) solution is added to an aqueous solution containing traces of a phenolic group, a color change occurs. The color varies from reds to blues to greens depending on the particular phenol molecule present.

Mole Relationships in a Chemical Reaction

The mole relationships in a balanced equation are used to make predictions about a reaction before it is performed in the laboratory, predictions such as: what mass of a product can be prepared from a known mass of a reactant?

The following example will illustrate how these predictions can be made for the reaction:

$$2\,C_2H_6 + 7\,O_2 \longrightarrow 4\,CO_2 + 6\,H_2O$$

How many grams of carbon dioxide will form if 4.150 grams of ethane are oxidized?

The balanced equation must be inspected to find the mole ratio of carbon dioxide to ethane: four moles of carbon dioxide are produced by two moles of ethane. The conversions are as follows:

$$\text{moles of ethane} = 4.150\text{ g}\left(\frac{1\text{ mole}}{30.08\text{ g}}\right) = 0.1380\text{ mole ethane}$$

$$\text{moles of } CO_2 = 0.1380\text{ mole ethane}\left(\frac{4\text{ moles } CO_2}{2\text{ moles ethane}}\right) = 0.2760\text{ mole } CO_2$$

$$\text{mass of } CO_2 = 0.2760\text{ mole}\left(\frac{44.01\text{ g}}{1\text{ mole}}\right) = 12.15\text{ grams } CO_2$$

The three steps needed to solve this problem were:

1) convert grams of ethane to moles of ethane,
2) convert moles of *ethane* to moles of *carbon dioxide using the balanced equation,*
3) convert moles of carbon dioxide to grams of carbon dioxide.

You will perform a calculation similar to this one; you will use the mole relationship in the balanced equation for the synthesis of aspirin to predict the mass of purified aspirin you should theoretically have obtained based on the mass of salicylic acid that you used as starting material.

Percentage Yield of Product

The amount of product obtained in a chemical reaction (experimental mass) is almost always less than what was expected (theoretical mass). This difference occurs because of incomplete reaction of the reactants, side reactions (in which different products are formed), and/or loss of product from careless experimental technique. The smaller the difference is between the experimental and theoretical masses, the more successful is the synthesis. The success is measured in terms of percentage yield, which is calculated using the formula:

$$\text{percent yield} = \left(\frac{\text{experimental mass}}{\text{theoretical mass}}\right) \times 100$$

Procedure that you will follow

A. Equipment Set-Up: Ice Bath

1. Partially fill a trough with ice (several students can share). Add about 100 mL of distilled water to an Erlenmeyer flask and place it into the ice bath to cool

B. Aspirin Synthesis

1. Place a *known* amount (close to 6 grams) of salicylic acid into a 125-mL Erlenmeyer flask.

2. Add about 12 mL of acetic anhydride and 10 drops of concentrated (18 M) sulfuric acid to the flask. Swirl the flask to mix the reactants. Allow the mixture to sit at room temperature for 15 minutes with occasional swirling.

During this time, assemble the vacuum filtration apparatus. Set-up and use of this apparatus is discussed on page 9 of this laboratory manual.

C. Separation of Product from the Reaction Mixture

1. Cool the reaction flask by placing it in the ice bath. Stir constantly; the mixture should thicken to a semi-solid sludge.

2. Add slowly, with stirring, 70 mL of the cooled water. Continue stirring until the mixture is a watery sludge. (Place the remaining cooled water back into the ice bath. It will be used during the vacuum filtration.) Allow the mixture to continue cooling in the ice bath for 10 to 15 minutes.

During this time, determine and record the mass of your watchglass and a piece of filter paper.

3. Vacuum filter the mixture. Use small portions of the cooled water to assist in rinsing the last of the crystals from the flask.

4. When filtration is complete, disconnect the vacuum tubing from the flask, then shut off the aspirator. Discard the solution in the filter flask. If all of the cooled water has been used, place another 50 mL of distilled water into the flask and return the flask to the ice bath. This water will be needed for the second vacuum filtration.

D. Recrystallization of the Product

1. Transfer the impure aspirin from the Buchner funnel to a 150-mL beaker. Add 25 mL of 95% ethanol.

2. Add about 50 mL of warm (50-70 °C) distilled water and stir to dissolve the impure aspirin. The mixture can be heated gently until the solid completely dissolves, but do not allow the mixture to boil.

3. Place the beaker in the ice bath to cool. Stir the mixture constantly for the first five minutes as crystals begin forming. Stir frequently after that, until the beaker has been in the ice bath for a total of about 15 minutes.

4. Place the pre-weighed filter paper in the Buchner funnel and vacuum filter the purified aspirin. Use small amounts of the cooled water to help rinse the last of the aspirin crystals from the beaker.

5. When filtration is complete, transfer the purified aspirin and filter paper from the Buchner funnel to the pre-weighed watchglass. Place a match head-sized quantity of the pure aspirin into a large test tube to be tested with iron (III) chloride as described in part E below. Spread the aspirin out on the watchglass to facilitate drying, and place the watchglass into the laboratory oven.

While the aspirin is drying, perform the iron (III) chloride test.

E. Iron (III) Chloride Test for Phenols

1. Into a second large test tube place a match head-sized quantity of salicylic acid. Add 5 mL of distilled water to this tube, to the tube containing the portion of your purified aspirin, and to an empty tube. This third tube will be used as a control for color comparison. Agitate the tubes to dissolve the solids as completely as possible. Add 2 to 3 drops of 0.2 M $FeCl_3$ solution to each tube and shake. An orange to purple color is indicative of the presence of the phenol group of salicylic acid. Record your observations.

F. Yield of Purified Aspirin

1. Shortly before the laboratory period ends, remove your purified aspirin from the oven and allow it to cool to room temperature. Determine and record the mass of the watchglass, filter paper, and dried aspirin.

2. Perform a mole relationship calculation to predict the mass of purified aspirin you should theoretically have obtained based on the mass of salicylic acid that you used as starting material. Show all of your work, complete with units, and express your answer to the appropriate number of significant figures. Compare the theoretical and experimental masses by calculating a percentage yield. If your technique has been good, you should have a high yield of pure aspirin.

DO NOT TAKE THE ASPIRIN HOME, AND DO NOT ATTEMPT TO INGEST ANY OF IT.

Report Sheet: Experiment 3 Synthesis of Aspirin

Name ______________________ *Date* __________ *Section* ________

Partner's Name ______________________ *Instructor's Initials* ______________

Aspirin Synthesis

Mass of salicylic acid ____________

Recrystallization of Product

Mass of watchglass ____________

Mass of filter paper ____________

Mass of watchglass, filter paper, and dried aspirin ____________

Mass of dried aspirin ____________

Results

Mass of aspirin expected ____________
Calculation:

Percentage yield of aspirin ____________
Calculation:

Iron (III) Chloride Test for Phenols

Salicylic Acid

Observation:

Conclusion:

Purified Aspirin

Observation:

Conclusion:

Blank

Observation:

Conclusion:

Additional Question:

What mass of salicylic acid should you start with if you wish to prepare 12.5 grams of methyl salicylate (oil of wintergreen) by the reaction shown on page 30? (Note that you will need to calculate the formula weight of methyl salicylate in order to perform this calculation.) Show all of your work, complete with units, and express your answer to the appropriate number of significant figures.

Experiment 4

Characterization of Carbohydrates

What you will accomplish in this experiment

You will use a variety of chemical tests to identify the special structural features of mono- and disaccharides. With these reactions, you'll be able to determine whether an unknown substance is a

- carbohydrate
- mono- or disaccharide
- ketose
- pentose
- reducing sugar.

Concepts to know to be prepared

Carbohydrates are ubiquitous compounds that are found in all living plant and animal organisms. As an aid to the study of these diverse compounds, a classification system is used:

- *monosaccharides* (simple sugars) consist of a single carbohydrate unit that does not hydrolyze (break apart with the addition of water) to a simpler carbohydrate;

- *oligosaccharides* (disaccharides, trisaccharides, etc.) are carbohydrates that hydrolyze to liberate the corresponding number of monosaccharide units;

- *polysaccharides* are polymers of monosaccharide units which when hydrolyzed yield monosaccharides. The number of monosaccharide units in a polysaccharide is usually greater than twelve and may reach several thousand.

Monosaccharides

The most common monosaccharides are pentoses (5 carbons) and hexoses (6 carbons). In their open-chain form, they contain an aldehyde or ketone group as well as several alcohol groups Thus they are polyhydroxyaldehydes or polyhydroxyketones and they exhibit chemical properties that are characteristic of these functional groups.

One consequence of these structures is the intramolecular reaction of an alcohol group with the carbonyl carbon of an aldehyde or ketone to form the cyclic hemiacetal or cyclic hemiketal, respectively.

The monosaccharides that will be investigated in this experiment are shown here in the open chain form as Fischer projections and in the cyclic form as Haworth projections:

D-ribose (aldopentose)

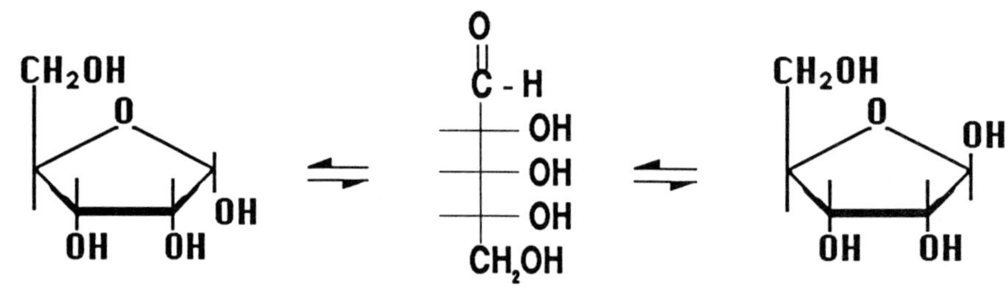

∂-anomer open-chain ß-anomer

D-glucose (aldohexose)

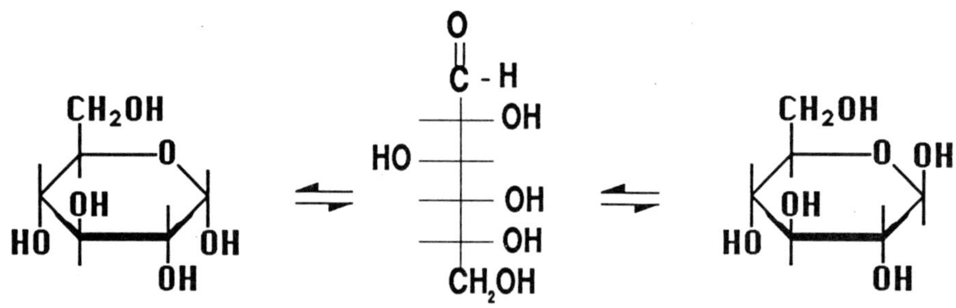

∂-anomer open-chain ß-anomer

D-fructose (ketohexose)

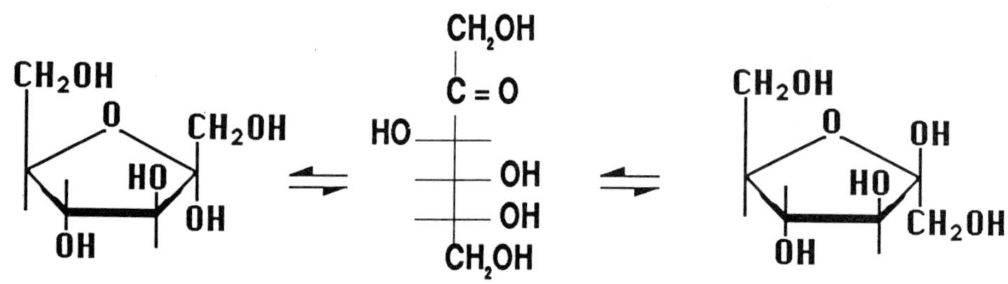

∂-anomer open-chain ß-anomer

When open-chain monosaccharides take the cyclic form, two stereoisomers are created: the alpha (∂) and beta (ß) anomers. These anomers differ only in the spatial arrangement of the groups at the anomeric carbon; that is, the carbon which would be the carbonyl carbon in the open-chain form. The anomer that is produced will depend on the orientation of the carbonyl at the time of ring closure. In solution, an equilibrium exists between the open-chain, alpha (∂), and beta (ß) forms. Glucose in an aqueous solution exists 99% in the cyclic form and only 1% in the open-chain form.

Carbohydrates with hemiacetals or hemiketal functional groups are called *reducing sugars* because their rings can open to give the free aldehyde or ß-hydroxyketone, which can then act as a reducing agent. The aldehyde or ß-hydroxyketone can reduce another participant in a reaction, and will in turn be oxidized.

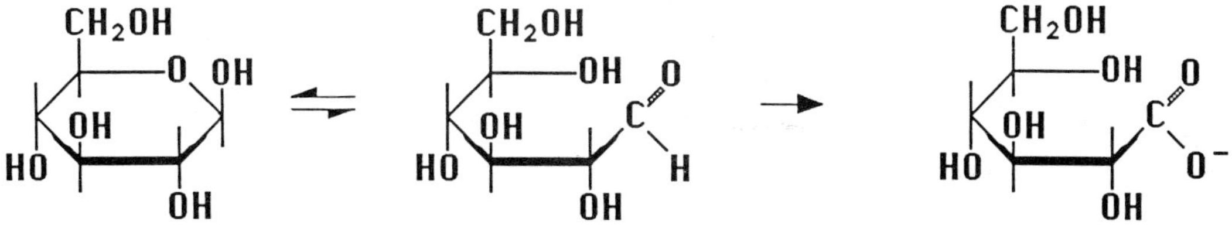

hemiacetal | open-chain aldehyde | oxidized form: carboxylic acid

Disaccharides

Disaccharides are formed by the reaction of a hemiacetal group on one monosaccharide unit with an alcohol group on another monosaccharide unit; the resulting functionality is an acetal. The bond connecting the monosaccharide units is commonly called a glycosidic linkage.

monosaccharide (glucose) | monosaccharide (glucose) | disaccharide (maltose)

This disaccharide, maltose, is a reducing sugar because it still contains a hemiacetal functional group:

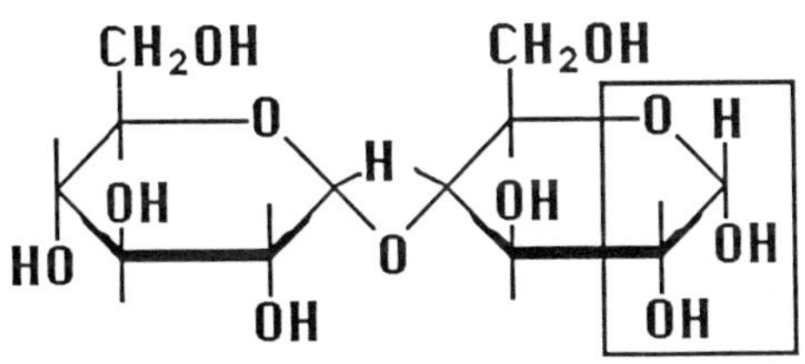

hemiacetal group

In basic solutions, this ring will open and the aldehyde will be restored and will be ready to serve as a reducing agent.

Compare the structure of maltose with that of another disaccharide, sucrose. Note that sucrose is a nonreducing sugar: there is no hemiacetal group. This is because the glucose and fructose subunits are joined through their anomeric positions; both anomeric carbons are involved in acetal functionalities. Acetal groups are stable under oxidizing conditions; they will not revert back to the free aldehyde or ß-hydroxyketone form, and there will be no oxidation reaction.

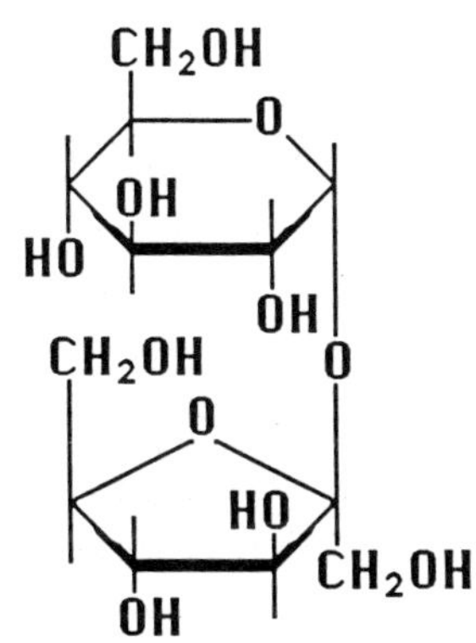

sucrose

Discussion of Chemical Tests

1. Dehydration of Monosaccharides

In the presence of a strong acid, monosaccharides undergo dehydration to yield cyclic alkenes called furfurals. Disaccharides are hydrolyzed by the acid to yield the monosaccharide components, which are dehydrated to the furfural.

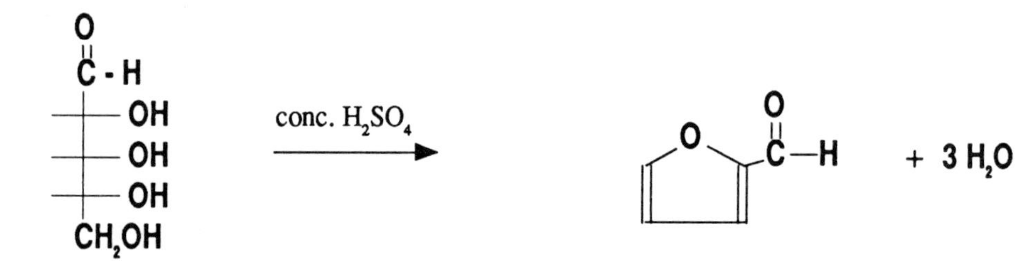

pentose furfural

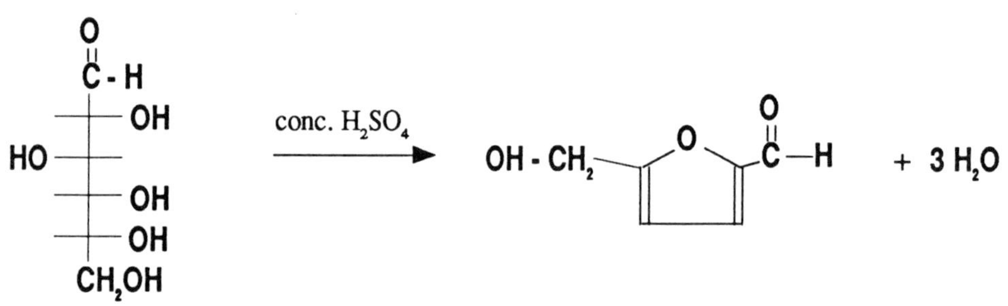

hexose 5-hydroxymethylfurfural

A. The Molisch Test: Diagnosis of Carbohydrates

The Molisch Test identifies any carbohydrate with five or more carbon atoms. All furfurals react with ∂-napthol to form a purple or violet compound.

B. Seliwanoff's Test: Diagnosis of Ketoses

The furfurals from ketoses react with resorcinol in minutes to give a positive Seliwanoff's Test. The color produced will depend on the number of carbons in the ketose: ketohexoses such as fructose give a cherry-red solution; ketopentoses give colors ranging from blue to green.

C. Bial's Test: Diagnosis of Pentoses

Bial's Test is used to distinguish pentoses from hexoses; this distinction is based on the color that develops in the presence of orcinol and iron (III) chloride. Furfural from pentoses gives a blue or green color. The furfural from hexoses may give a muddy-brown or gray solution, but this is easily distinguishable from the green color of pentoses.

2. Reduction of Sugars

A. Benedict's Test: Diagnosis of Reducing Sugars

Benedict's reagent is a basic solution containing Cu^{+2} ions complexed with the polyatomic citrate ion $C_6H_6O_7^{-2}$. Solution of Cu^{+2} ions are characteristically blue. Monosaccharides and disaccharides that contain reducing groups will reduce the Cu^{+2} ions to Cu^{+1}, which will precipitate as the brick-red copper (I) oxide, Cu_2O.

B. Barfoed's Test: Diagnosis of Monosaccharides

As is true of the test just described, Barfoed's Test depends on the reducing properties of the saccharides being tested. However, the acidic test conditions allow for selectivity between monosaccharides and disaccharides. Monosaccharides yield the brick-red Cu_2O precipitate within two or three minutes. Disaccharides require ten minutes or more to produce the red precipitate.

Procedure that you will follow

The following procedures are to be carried out on samples of known carbohydrates and on an unknown carbohydrate. The unknown should be tested at the same time as the known sugars to insure that each compound is tested under the same conditions. Label all test tubes to avoid confusion.

A. The Molisch Test: Diagnosis of Carbohydrates

Compounds to be tested are 1% solutions of ribose, glucose, fructose, maltose, sucrose, and the unknown.

1. Place 2 mL of the sugar solution and 2 drops of Molisch reagent in a large test tube (use a separate test tube for each sugar).

2. Incline the tube, and with a pipet, carefully add approximately 1-2 mL of concentrated sulfuric acid so that it flows down the side of the tube and forms a layer beneath the aqueous layer. Note any colors form at the interface of the layers after a short time (2 to 3 minutes). A green ring may form initially due to impurities in the sulfuric acid. A positive test for carbohydrates is the slower formation of a red-violet layer at the interface.

B. Barfoed's Test: Diagnosis of Monosaccharides

Compounds to be tested are 1% solutions of ribose, glucose, fructose, maltose, sucrose, and the unknown.

1. Place 5 mL of Barfoed's reagent and 5 mL of the sugar solution in a large test tube (use a separate test tube for each sugar). Mix each tube well.

2. Place all of the test tubes at the same time in an actively boiling water bath. Heat for 3.5 minutes after the water starts boiling again. Timing is important, as a positive test can be obtained with disaccharides if heated longer than 3.5 minutes. During this period, observe the tubes closely and note any change in color or clarity of the solutions. A red precipitate of copper(I) oxide is a positive test for monosaccharide.

C. Seliwanoff's Test: Diagnosis of Ketoses

Compounds to be tested are 1% solutions of ribose, glucose, fructose, maltose, sucrose, and the unknown.

1. Place 5 mL of Seliwanoff's reagent and 2 drops of sugar solution in a large test tube (use a separate test tube for each sugar). Mix each tube well.

2. Place all of the test tubes at the same time in an actively boiling water bath. Record the color you observe in each tube after 4 minutes. Ketohexoses produce cherry-red solutions, and ketopentoses give blue-to-green solutions. The various colors of other carbohydrates constitute a negative test. (Aldohexoses may produce a light yellow to faintly pink color.)

D. Bial's Test: Diagnosis of Pentoses

Compounds to be tested are 1% solutions of ribose, glucose, fructose, maltose, sucrose, and the unknown.

1. Place 3 mL of Bial's reagent and one drop of sugar solution in a large test tube (use a separate test tube for each sugar). Mix each tube well.

2. Place all of the test tubes at the same time in an actively boiling water bath and heat until color develops. A pentose will produce a green (or blue) color within 2-3 minutes. The various colors of other carbohydrates constitute a negative test. (Ketoses or disaccharides may produce a muddy-brown or gray color in about two minutes; aldohexoses may produce a brown color only after heating for several more minutes.)

E. Benedict's Test: Diagnosis of Reducing Sugars

Compounds to be tested are 1% solutions of ribose, glucose, fructose, maltose, sucrose, and the unknown.

1. Place 5 mL of Benedict's reagent and 1 mL of sugar solution in a large test tube (use a separate test tube for each sugar). Mix each tube well.

2. Place all of the test tubes at the same time in an actively boiling water bath. Heat for 5-6 minutes. The formation of a red copper (I) oxide precipitate is a positive test for reducing sugars.

Report Sheet: Experiment 4
Characterization of Carbohydrates

Name ______________________ Date __________ Section ________

Partner's Name ______________________ Instructor's Initials ______________

1. The Molisch Test: Diagnosis of Carbohydrates

Compound Tested	Observations	Conclusions
Ribose		
Glucose		
Fructose		
Sucrose		
Maltose		
Unknown		

2. *Barfoed's Test: Diagnosis of Monosaccharides*

Compound Tested	Observations	Conclusions
Ribose		
Glucose		
Fructose		
Sucrose		
Maltose		
Unknown		

3. *Seliwanoff's Test: Diagnosis of Ketoses*

Compound Tested	Observations	Conclusions
Ribose		
Glucose		
Fructose		
Sucrose		
Maltose		
Unknown		

4. Bial's Test: Diagnosis of Pentoses

Compound Tested	Observations	Conclusions
Ribose		
Glucose		
Fructose		
Sucrose		
Maltose		
Unknown		

5. Benedict's Test: Diagnosis of Reducing Sugars

Compound Tested	Observations	Conclusions
Ribose		
Glucose		
Fructose		
Sucrose		
Maltose		
Unknown		

Characterization of Unknown Carbohydrate Number: ___________

Your unknown carbohydrate is one of the following carbohydrates:

ribose, glucose, fructose, maltose, sucrose.

Which of the five possible carbohydrates is your unknown? ______________________________

Write a short paragraph summarizing the test results for your unknown and explaining how you were able to deduce its identity.

Experiment 5

Properties and Preparation of Esters and Soaps

What you will accomplish in this experiment

You will synthesize soap by saponification of a triacylglycerol, and then study the properties of the soap in water. You will also prepare several esters by condensation of various acids and alcohols.

Concepts to know to be prepared

Fatty Acids

Fatty acids are long, straight-chain carboxylic acids; those found in nature usually contain an even number of carbon atoms. Fatty acids with only single bonds between carbon atoms are said to be *saturated.* *Unsaturated* fatty acids have one or more double bonds between carbon atoms. The common names of fatty acids are used more frequently than the IUPAC names.

Some Saturated Fatty Acids

name	formula
Butyric Acid	$CH_3(CH_2)_2COOH$
Lauric Acid	$CH_3(CH_2)_{10}COOH$
Myristic Acid	$CH_3(CH_2)_{12}COOH$
Palmitic Acid	$CH_3(CH_2)_{14}COOH$
Stearic Acid	$CH_3(CH_2)_{16}COOH$
Arachidic Acid	$CH_3(CH_2)_{18}COOH$

Some Unsaturated Fatty Acids

name	formula
Oleic Acid	$CH_3(CH_2)_7CH{=}CH(CH_2)_7COOH$
Linoleic Acid	$CH_3(CH_2)_4CH{=}CHCH_2CH{=}CH(CH_2)_7COOH$
Linolinic Acid	$CH_3CH_2CH{=}CHCH_2CH{=}CHCH_2CH{=}CH(CH_2)_7COOH$
Arachidonic Acid	$CH_3(CH_2)_4CH{=}CHCH_2CH{=}CHCH_2CH{=}CHCH_2CH{=}CH(CH_2)_3COOH$

Fatty acids, like all carboxylic acids, will readily react with bases to form carboxylates, the salts of the fatty acids.

$$\underset{\text{carboxylic acid}}{RCOOH} + \underset{\text{base}}{NaOH} \longrightarrow \underset{\text{salt}}{RCOO^- \; Na^+} + \underset{\text{water}}{H_2O}$$

Some examples of salts of acids are shown below.

$$\underset{\text{sodium acetate}}{CH_3COO^- \; Na^+} \qquad \underset{\text{potassium palmitate}}{CH_3(CH_2)_{14}COO^- \; K^+}$$

Esters

The general formula for an ester is $R-\overset{\overset{\displaystyle O}{\|}}{C}-O-R'$, where R is the alkyl group of the carboxylic acid, and R' is the alkyl group of the alcohol from which the ester has been prepared. The condensation reaction is shown below:

$$\underset{\text{carboxylic acid}}{R-\overset{\overset{\displaystyle O}{\|}}{C}-O-H} + \underset{\text{alcohol}}{H-O-R'} \xrightarrow{H_2SO_4} \underset{\text{ester}}{R-\overset{\overset{\displaystyle O}{\|}}{C}-O-R'} + \underset{\text{water}}{H_2O}$$

Esters are named as binary compounds: the first word of their name comes from the alkyl group of the alcohol, while the second word is formed by changing the "-ic" ending of the acid from which they originate to "-ate."

$$\underset{\text{methyl acetate}}{CH_3-\overset{\overset{\displaystyle O}{\|}}{C}-O-CH_3} \qquad \underset{\text{pentyl acetate}}{CH_3-\overset{\overset{\displaystyle O}{\|}}{C}-O-CH_2CH_2CH_2CH_2CH_3}$$

Many esters are colorless liquids with pleasant, fruity odors; esters in fruits and flowers and are responsible for much of their flavor and aroma.

Triacylglycerols

Triacylglycerols are fatty acid triesters of the alcohol glycerol.

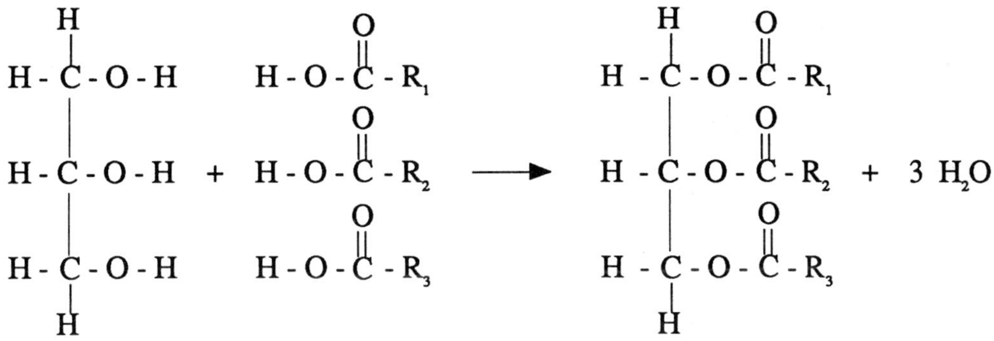

glycerol fatty acids triacylglycerol water

R_1, R_2, and R_3 represent fatty acid chains which may or may not be the same length.

The structure on the left below is *glyceryl tristearate, a fat* that is present in lard. Fats are composed of high proportions of high molecular weight *saturated* fatty acids. The structure on the right below is *glyceryl trioleate,* an *oil* that is present in corn oil. Oils have a relatively high proportion of high molecular weight *unsaturated* fatty acids.

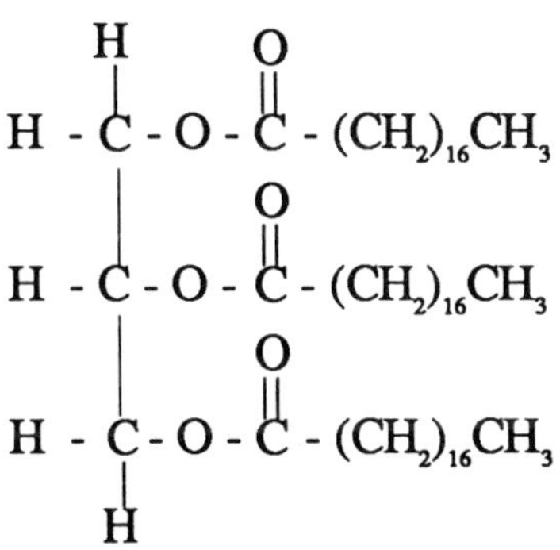

glyceryl tristearate

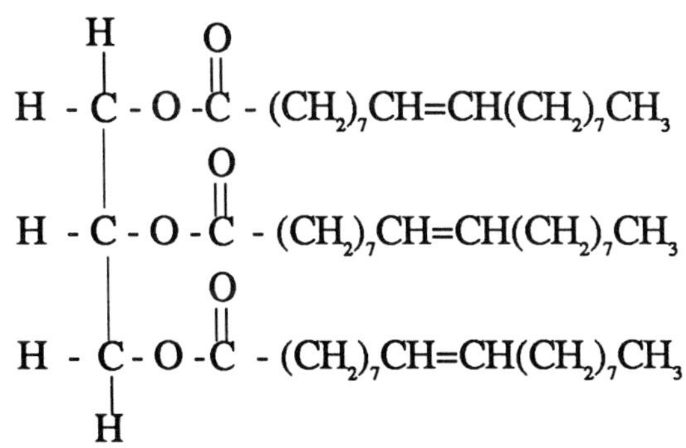

glyceryl trioleate

Hydroysis of Triacyglycerols

Fats and oils can be hydrolyzed to glycerol and the constituent fatty acids by superheated steam or by certain lipase enzymes of the digestive tract:

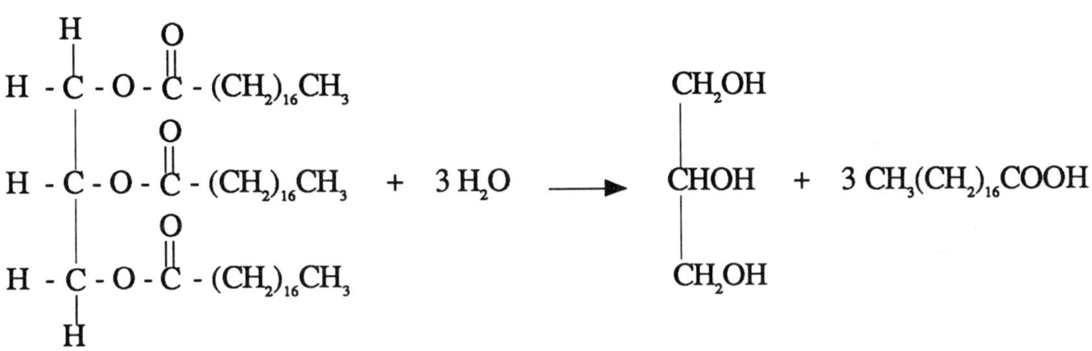

If the hydrolysis is takes place under basic conditions, it is called a *saponification,* and the products are glycerol and the *salt* of the fatty acid. *The sodium or potassium salt of a fatty acid is a soap.*

$$\text{glyceryl tristearate} + 3\,NaOH \longrightarrow \text{glycerol} + 3\ \text{sodium stearate}$$

Sodium salts are the ordinary hard soaps; potassium soaps are soft. Ordinary sodium soaps are not very soluble, although we tend to think of them that way. They are easily precipitated ("salted out") from solution by high concentrations of salt.

The Cleansing Action of Soaps

A soap molecule has a large nonpolar hydrocarbon portion and a carboxylic salt end:

$$CH_3CH_2CH_2CH_2CH_2CH_2CH_2CH_2CH_2CH_2CH_2CH_2CH_2CH_2CH_2-\overset{\overset{\displaystyle O}{\|}}{C}-O^-\ Na^+$$

hydrophobic (repelled by water) — hydrophilic (water soluble)

When soap is added to water, the hydrophilic ends of the molecules are attracted to the water and dissolve in it. The hydrophobic ends are repelled by the water molecules. Consequently, a thin film of soap (suds) forms on the water surface, drastically lowering the surface tension.

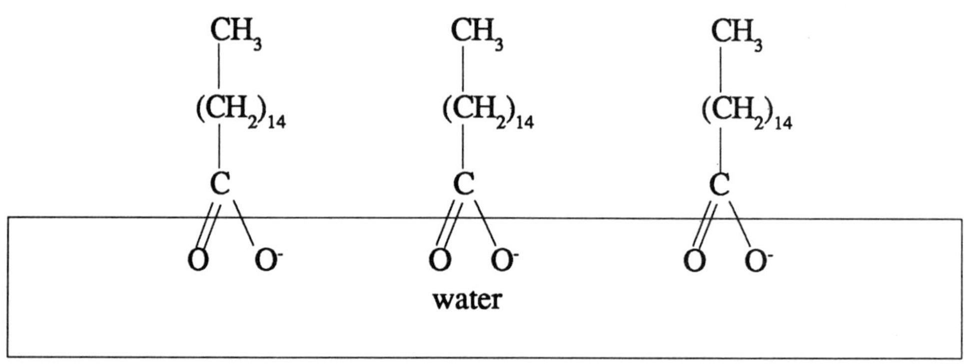

When a soap solution is brought into contact with oil (most dirt is held to clothes by a thin film of oil), soap molecules become reoriented: the hydrophobic portions dissolve in the oil, and the hydrophilic ends remain dissolved in the aqueous phase. Mechanical action (scrubbing) causes the oil to disperse into tiny droplets, and the soap molecules arrange themselves around the surface of the globules to form micelles. Because the carboxylate ends of the soap molecules project outward, the surface of the micelle is negatively charged; the drops repel one another and do not coalesce (form larger drops). The entire micelle becomes water soluble and can be washed away by a strean of water. The cleansing process involves both the lowering of the surface tension of water and the production of micelles.

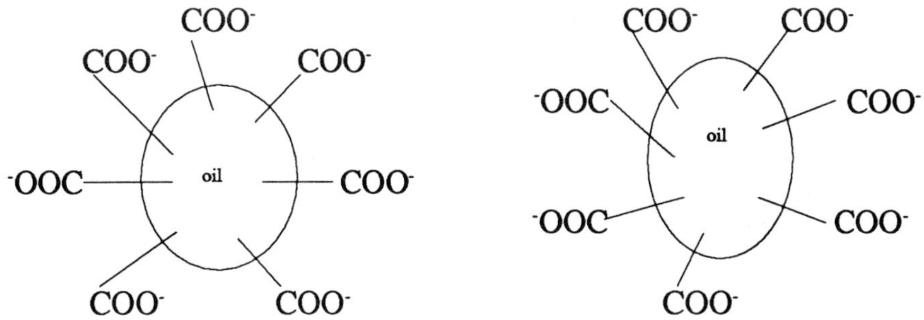

Certain metal ions present in "hard water" decrease the cleansing power of soaps. Calcium and magnesium ions will form precipitates with the carboxylate ions of fatty acids. These precipitates are responsible for the bathtub ring.

$$2\ CH_3(CH_2)_{16}\overset{\overset{\displaystyle O}{\|}}{C}\text{-}O^- + 2\ Na^+ + Ca^{+2} \longrightarrow [CH_3(CH_2)_{16}\overset{\overset{\displaystyle O}{\|}}{C}\text{-}O^-]_2\ Ca^{+2} + 2\ Na^+$$

The various methods for softening hard water all involve the removal of the calcium and magnesium ions.

Procedure that you will follow

A. Preparation of Some Esters by Condensation

1. Heat about 300 mL of tap water in an 800-mL beaker to 85°C. While the water is heating, label four large test tubes (A, B, C, and D) and obtain corks for each. In the respective tubes place the listed ingredients:

Tube A
Add 20 drops of ethanol, 20 drops of acetic acid (ethanoic acid), and 10 drops of concentrated sulfuric acid.

Tube B
Add a quantity of salicylic acid equivalent to the size of a match head, 20 drops of methanol, and 3 drops of concentrated sulfuric acid.

Tube C
Add 20 drops of pentanol, 20 drops of acetic acid (ethanoic acid), and 5 drops of concentrated sulfuric acid.

Tube D
Add a quantity of benzoic acid equivalent to the size of a match head, 20 drops of ethanol, and 5 drops of concentrated sulfuric acid.

Stir each test tube with a stirring rod, rinsing the rod after each tube is stirred, and place a cork stopper *loosely* in each.

2. When the water bath has reached 85°C, turn off the burner and place each stoppered flask in the water bath. After several minutes, sample the odor of each flask. You should be able to detect the odor of banana oil, butter rum, fingernail polish remover, and wintergreen. Record the odor of each mixture.

B. Preparation of a Soap by Saponification of a Triacylglycerol

1. Place 23 mL of cottonseed oil into a 250-mL Erlenmeyer flask. Add 20 mL of ethanol (as a solvent) and 20 mL of 25% sodium hydroxide solution.

2. Gently heat the flask in a boiling water bath, stirring the mixture frequently. After heating for twenty minutes, the odor of the ethanol should disappear, indicating the completion of the reaction. A pasty mass containing a mixture of the soap, glycerol, and excess sodium hydroxide is obtained.

3. Use an ice-water bath to cool the flask and its contents.

C. Investigation of the Properties of Soap

Part 1. Separation of the Soap by Salting-Out
To precipitate or "salt out" the soap, add 150 mL of a saturated sodium chloride solution to the soap mixture while stirring. The soap will float to the top of the mixture. Isolate the soap by vacuum filtration, washing it with ice-cold water. (Vacuum filtration is described on page 9 of this laboratory manual.) Record your observations of the soap. Attempt to form suds by shaking a small piece of the soap with water in a large test tube. Record your observations.

Part 2. Effect of Hard Water on Soap
Place a small amount (~1/8 teaspoon) of the soap you have prepared in a 50-mL beaker containing 25 mL of water. Warm the beaker and its contents to dissolve the soap. Pour 5 mL of the soap solution into each of four large test tubes. Add "hard water" solutions to three of the tubes as described below: the fourth tube will be used in part 3.

Tube 1
Add several drops of 0.2 M calcium chloride solution.

Tube 2
Add several drops of 0.2 M magnesium chloride solution.

Tube 3
Add several drops of 0.2 M lead acetate solution.

Record your observations for each test tube.

Repeat this procedure using 5 ml portions of a clear, 5-10% solution of a synthetic detergent.

Part 3. Formation of the Fatty Acid from the Carboxylate
To the fourth test tube, add 6 M HCl, one drop at a time, until a precipitate forms. Record your observations.

Report Sheet: Experiment 5
Properties and Preparation of Esters and Soaps

Name ______________________________ *Date* ____________ *Section* __________

Partner's Name ______________________________ *Instructor's Initials* ________________

A. Preparation of Some Esters by Condensation

Tube A. Reaction of acetic acid (ethanoic acid) and ethanol.

Describe the odor.

Give the equation for the reaction and the names of the products.

Tube B. Reaction of salicylic acid and methanol.

Describe the odor.

Give the equation for the reaction and the names of the products.

Tube C. Reaction of acetic acid (ethanoic acid) and 1-pentanol.

Describe the odor.

Give the equation for the reaction and the names of the products.

Tube D. Reaction of benzoic acid and ethanol.

Describe the odor.

Give the equation for the reaction and the names of the products.

B. Preparation of Soap by Saponification of a Triacylglycerol

Describe the appearance of the soap.

Give the equation for the reaction of sodium hydroxide and glyceryl trioleate.
Give the names of all products.

C. Investigation of the Properties of Soap

Part 1. Separation of the Soap by Salting Out

Describe the appearance of the resulting material.

Did the collected soap form suds with distilled water?

Question: This salting out process is used commercially. Using this process, what substance could be separated from the prepared soap? (Consider the formation of the soap by saponification.)

Part 2: Effect of Hard Water on Soap

Describe the result when solutions of Ca^{+2}, Mg^{+2}, and Pb^{+2} are added to:

- solutions of your soap.

- solutions of a synthetic detergent.

Give the equation for the reaction of $CaCl_2$ with the soap, sodium oleate. Be sure to balance the equation.

Part 3: Formation of Fatty Acid from the Carboxylate

Describe the appearance of the resulting material.

Write the equation of the reaction of the soap, sodium oleate, with HCl. Give the names of the products.

Experiment 6

Paper Chromatography of Amino Acids

What you will accomplish in this experiment

You will utilize the technique of paper chromatography to separate and identify amino acids.

Concepts to know to be prepared

Amino Acids

Amino acids are the basic structural units of proteins. Each amino acid consists of a central carbon atom to which four distinct groups are bonded: an amino group, a carboxylic acid group, a hydrogen atom, and an R group (side chain). Twenty sides chains varying in size, shape, charge, chemical reactivity, and hydrogen bonding capacity are commonly found in proteins. The amino acids in a mixture can be separated and identified on the basis of the different properties of these side chains; a technique that can be used to achieve this separation is paper chromatography.

Paper Chromatography

Chromatography is a means of separating the components of a mixture; the separation is achieved by allowing the components (solute) to interact with and distribute between two phases, a stationary phase (sorbent) and a mobile phase (solvent).

In paper chromatography the sorbent is a layer of water bound to a cellulose matrix. The solutes are applied at one end of the paper and then allowed to dry. The end of the paper near the application point (origin) is then placed in contact with a mixture of solvents which are to serve as the mobile phase. This phase moves through the paper by capillary action. Separation of components in the spotted sample will occur as the solutes interact to different degrees with the stationary water layer and the moving solvent phase. Some of the solute particles will spend more time in the mobile phase because of their greater solubility in that phase; these moleculeswill progress further along the paper than those that are less soluble. The molecules with a greater solubility in the stationary phase will not progress very far along the paper.

When the solvent has progressed most of the way up the paper, the paper is removed from the solvent, the solvent front is marked, and the paper is allowed to dry. The solutes are visualized by conversion to a brightly colored derivative; amino acids and peptides react with ninhydrin to give purple compounds. The visible components can then be identified by comparing their R_f values to those of standard, known compounds. The R_f value is an expression of the flow rate of a type of particle with respect to the flow rate of the mobile phase is expressed:

$$R_f = \frac{\text{distance solute moved from origin}}{\text{distance solvent moved from origin}}$$

In this experiment, you will determine the R_f values of eight amino acids. You will be given an unknown containing one or two of these eight amino acids. You will identify your unknown(s) by comparing their R_f values to those of the known amino acids.

Procedure that you will follow

1. Place 30 mL of a solution consisting of isopropanol, water, and concentrated ammonia (40:20:1 v/v/v) in a one liter beaker. Cover the beaker with a sheet of aluminum foil; crimp the foil down tightly over the mouth of the beaker. This beaker will serve as the chromatographic chamber, and it must be prepared at the start of the laboratory period so that the solution can equilibrate in the closed container.

2. Obtain a clean sheet of 13 cm by 23 cm Whatman #1 filter paper. Be careful to handle the filter paper *by the edges only* throughout the experiment.

3. Use a pencil and ruler to draw a line along the length of the paper, about 2 cm from the edge of the paper. Make nine x's along this line at intervals of 2 cm. Identify each x as illustrated below.

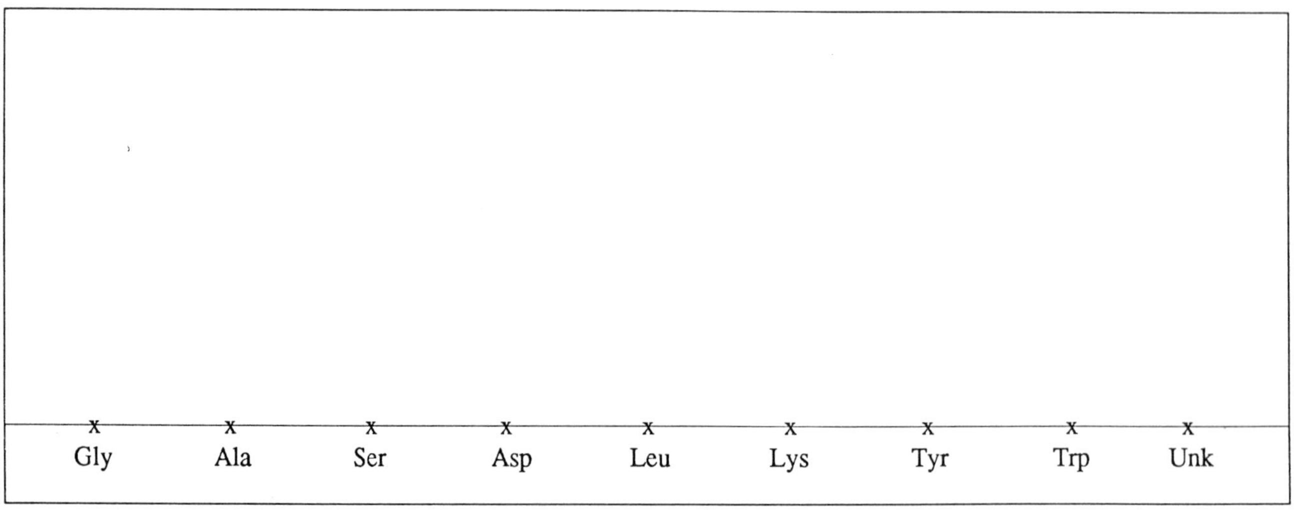

4. The following amino acids will be used as knowns: Glycine (Gly), Alanine (Ala), Serine (Ser), Aspartate (Asp), Leucine (Leu), Lysine (Lys), Tyrosine (Tyr), and Tryptophan (Trp). These will be provided as aqueous solutions containing 1% concentrated hydrochloric acid to promote complete dissolution of the amino acid. You must also obtain an unknown amino acid solution from your instructor. Use glass capillaries to spot each solution on the respective mark. Separate capillaries must be used for each solution, and the spots must be kept as small as possible (no more than 3 mm in diameter.) Each amino acid should be spotted twice, and the spots must be allowed to dry completely after each application.

5. When all spots are dry, roll the filter paper into a large cylinder, and staple the ends together; *the stapled ends must not overlap.*

6. Place the rolled paper, with the spots at the bottom, into the one liter beaker, and quickly replace the aluminum foil cover.

7. Allow the solvent to move up the paper until it is within one centimeter of the top. Remove the paper, unroll it, and use a pencil to mark the edge of the solvent front before the paper drys.

8. When the paper has dried, spray it with ninhydrin reagent, and place it in a 110 °C oven for five minutes to allow the color of the spots to develop.

9. Use a pencil to outline and mark the center of each spot; do this quickly, as the color of the spots will fade with time.

10. Determine the R_f values of the knowns and your unknown amino acid. Determine the identity of your unknown amino acid.

Report Sheet: Experiment 6
Paper Chromatography of Amino Acids

Name ______________________ *Date* __________ *Section* ________

Partner's Name ______________________ *Instructor's Initials* ______________

Determination of R_f values of Known Amino Acids

amino acid	distance solute moved from origin	distance solvent moved from origin	R_f value
glycine			
alanine			
serine			
aspartate			
leucine			
lysine			
tyrosine			
tryptophan			

Identification of Unknown Amino Acid(s)

identity	distance solute moved from origin	distance solvent moved from origin	R_f value

Additional Questions:

1. Compare your R_f values for the amino acids serine, aspartate, and leucine. Explain *on the molecular level* the reason for the differences in the rates of movement of these amino acids.

2. The R_f values of two amino acids are 0.40 and 0.60. What distance will each of these amino acids travel on a paper chromatogram is the solvent moved a distance of 10.0 cm?

Will these two amino acids be effectively separated on the chromatogram? Explain.

Will two amino acids with R_f values of 0.39 and 0.40 be effectively separated on the same chromatogram? Explain.

Experiment 7

Hydrolysis of Starch by Salivary Amylase

What you will accomplish in this experiment

Amylase, an enzyme present in saliva, catalyzes the hydrolysis of the glycosidic linkages in starch. The effect of enzyme concentration, substrate concentration, temperature, pH, and heavy metal cations on the activity of this enzyme will be investigated.

Concepts to know to be prepared

Metabolism

The chemical reactions that go on in living organisms, and which are essential for growth, reproduction, movement, and all other vital functions, are individually and collectively referred to as *metabolism*. Many thousands of such metabolic processes are occurring constantly in even the simplest forms of life.

Enzymes are proteins that act as catalysts for metabolic reactions. They increase the rate of the reaction, but do not influence the kind or amount of products formed. In general, each metabolic reaction has to be catalyzed in the living organism by its own special enzyme.

The existence of enzymes in biological materials can be demonstrated by the effects they bring about. In this experiment, digestive enzymes capable of breaking down such polysaccharides as starch will be studied.

Food materials *must* be broken down by digestion before they can be absorbed and utilized by the body. Consequently, *if an organism has no enzyme capable of attacking a particular type of carbohydrate, protein, or fat, that particular substance will have no food value for the organism.* Carbohydrates and fats can be broken down by strong acids or bases. However, digestive enzymes can accomplish the same breakdown under conditions compatible with life, that is, at moderate temperatures and physiological pH ranges.

Starch

The nutritional reservior in plants is starch, which is actually a mixture of two polysaccharides. *Amylose,* the *unbranched* type of starch, consists of glucose residues in ∂-1,4 linkage.

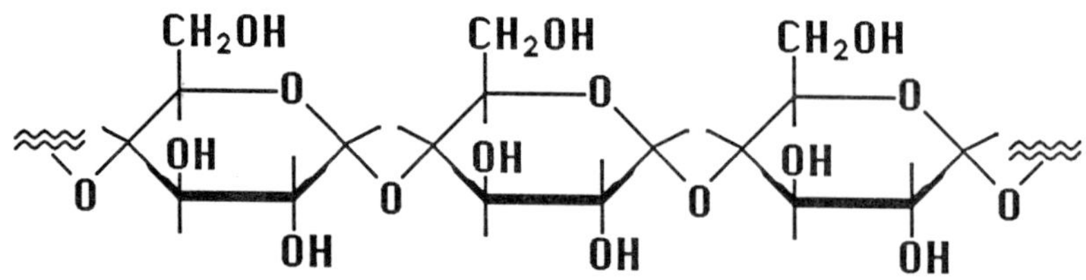

Amylopectin, the *branched* form, has about one ∂-1,6 linkage per thirty ∂-1,4 linkages.

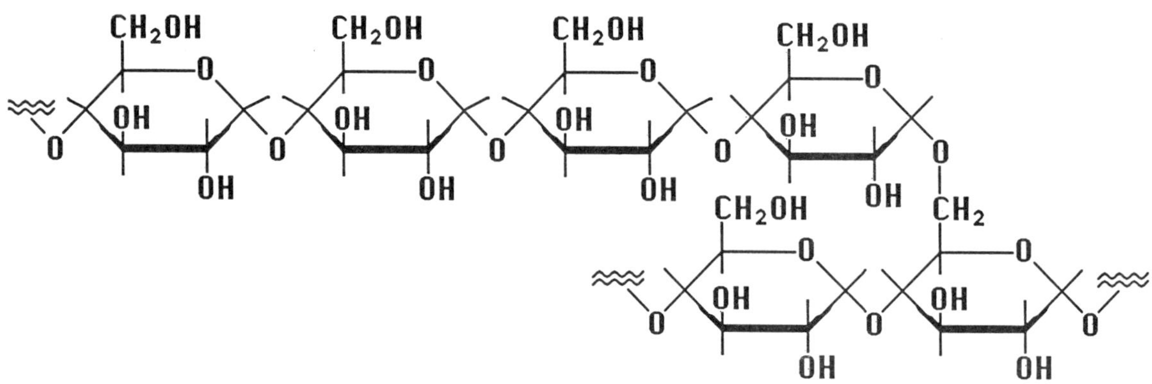

Amylases

Hydrolysis of starch is the process of digestion. Enzymes called *amylases* catalyze only the hydrolysis of ∂-1,4 glycosidic linkages in the amylose and amylopectin components of starch. The hydrolysis does not proceed directly from polysaccharides to monomer units; rather, partial hydrolysis products of intermediate size are obtained. These products are maltose and dextrin. Maltose consists of two glucose units in ∂-1,4 linkage; dextrin is made up of several glucose units joined by ∂-1,6 linkage in addition to ∂-1,4 linkages.

Amylase enzymes are present in saliva, so the digestion of carbohydrates begins in the mouth. Digestion continues briefly in the stomach until the pH drops too low, and then is completed in the intestines by the attack of another amylase.

The chemical conversion of starch to dextrins and maltose is a key part of the brewing step in the production of beer. The amylase enzymes are added to the grain mixture as brewer's malt. Approximately one third of the calories in regular beer are contributed by carbohydrate remaining in the beer after fermentation. The carbohydrates remain because amylase is unable to break down the ∂-1,6 glycosidic linkages in the amylopectin portion of the starch. To make "light beer," less starch is used in the inital brewing step and a second enzyme, amyloglucosidase, which *can* cleave the ∂-1,6 glycosidic linkage in amylopectin, is added in the fermentation step. The light beer will have a lower carbohydrate composition because of the added amyloglucosidase, and a lower alcohol content because of the reduced starch in the initial fermentation mixture.

Study of Enzyme Activity

To follow the action of an enzyme, it is necessary to test for the appearance of a product or the disappearance of a reactant over a measured period of time. In the reactions of amylase with starch, you will test for the disappearance of starch by reacting samples of the reaction mixtures with iodine. Initially, a starch solution gives a blue-black color with iodine. The amylase catalyzes the hydrolysis of the ∂-1,4 glycosidic linkages, forming maltose and dextrin. After the starch is hydrolyzed, the blue-black color produced with iodine is no longer observed, and only the red or gold coor of the iodine solution is seen. The faster the amylase hydrolyzes the starch, the more quickly the blue-black color disappears. If the blue-black color does not fade, it can be concluded that the enzyme is no longer active and that no hydrolysis of starch has occurred.

Enzyme activity depends on several factors, including enzyme concentration, substrate concentration, pH, temperature, and presence of heavy metal cations. Your own saliva containing the amylase enzyme will be used for this experiment, although the levels of amylase vary considerably from one person to another. Each experiment must be timed. As you proceed with each experiment, you will check enzyme activity by reacting a few drops of the reaction mixture with iodine. The time at which the blue-black color of starch disappears will be noted in each experiment.

The time required for the hydrolysis of starch will be correlated to the relative enzyme activity. When enzyme activity is high, the time for the starch to hydrolyze will be very short. When the enzyme is operating poorly or not at all, the activity is low, and more time will be required for the starch to hydrolyze. In some cases, the enzyme will be completely inactivated and the blue-black color of the starch and iodine will persist throughout the entire experiment. Graphs will be prepared showing the effects of concentration, pH, temperature, and heavy metal cations on the relative enzyme activity.

Procedure that you will follow

A. Effect of Enzyme Concentration

1. Collect approximately 5 mL of saliva in a small, clean beaker.

2. Place 5 mL of 1% starch solution in each of 5 large test tubes.

3. Place the tubes in a 37-40 °C water bath. After 5 minutes, add the following amounts of saliva to the test tubes as quickly as possible, mixing each solution thoroughly, and then returning the tubes to the water bath. *Do not overheat the bath or you will inactivate the enzyme.*

test tube	number of drops of saliva
1	1
2	2
3	5
4	10
5	20

4. Prepare a spot plate for testing for the presence of starch in the samples by placing one drop of iodine reagent in each of five depressions on the spot plate.

5. Two minutes after the addition of saliva, transfer one drop from each test tube (using a different pipet for each tube) to a separate drop of iodine in the spot plate. Note the color produced for each. *Remember*

that the complex formed by starch and iodine is an indigo blue. If the solor with iodine is red or gold, the starch has been completely hydrolyzed.

6. Clean the spot plate and then prepare it for the next testing by placing one drop of iodine reagent in each of five depressions.

7. Repeat the testing at 5 minutes, and at 5 minute intervals thereafter. Continue testing for 30 minutes, or until the blue-black color no longer appears for each sample.

8. Record the time required for the hydrolysis of starch in each sample.

9. Apply Benedict's and Barfoed's tests (these are discussed below) to 1 mL portions of any one of the hydrolyzed starch solutions, and also to 1 mL portions of *fresh* starch solution, so that you can make a comparison of the results of the tests.

B. Benedict's Test for Easily Oxidized Groups

The Benedict's reagent is a mild oxidizing agent consisting of the Cu^{+2} ion and the polyatomic citrate ion ($C_6H_6O_7^{-2}$) in basic solution. The Cu^{+2} forms a complex with citrate; it is wrapped by and protected within the citrate ion. This protection allows Cu^{+2} to remain dissolved in the basic solution (otherwise, the extremely insoluble $Cu(OH)_2$ would form). When an easily oxidized compound is added to Benedict's reagent, and the solution is warmed, Cu^{+2} is reduced to Cu^{+1}. These ions cannot be protected from OH^- by citrate, and precipitate as copper (I) oxide (Cu_2O). The Benedict's reagent has a bright blue color caused by the Cu^{+2} ion, but Cu_2O is a brick-red solid. Visible evidence of a positive Benedict's test is the disappearance of the blue color of the solution, and the appearance of a brick-red precipitate. Monosaccharides, and disaccharides that contain aldehydes or ß-hydroxy ketones will give a positive Benedict's test, as these groups are easily oxidized to carboxylic acids.

Carbohydrates to be tested: hydrolyzed starch and fresh starch solutions.

1. Place 5 mL of Benedict's reagent and 1 mL of starch solution in a large test tube (use a separate tube for each starch solution).

2. Shake the contents of each tube well, and place both tubes at the same time in an actively boiling water bath. Heat in the water bath for 3 minutes. On the report sheet, record any changes in the colors or transparencies of the solutions, and in the formation and color of precipitates.

C. Barfoed's Test for Monosaccharides

As is true of the Benedict's test, the Barfoed's test depends on the presence of easily oxidized groups to reduce Cu^{+2} to Cu^{+1} to form the brick-red precipitate of Cu_2O. However, the Barfoed's reagent is slightly acidic rather than basic; this difference allows for *selectivity* between the monosaccarides and disaccharides. Monosaccharides will test positive within two to three minutes. Disaccharides with oxidizable groups require ten minutes or more before they show the positive result.

Carbohydrates to be tested: hydrolyzed starch and fresh starch solutions.

1. Place 1 mL of the starch solution and 3 mL of Barfoed's reagent in a large test tube (use a separate tube for each starch). Shake the contents of each tube well, and place both tubes at the same time into an actively

boiling water bath. Heat in the water bath for 5 minutes *(no longer)*. During this period, observe the tubes closely and record any change in color or clarity of the solutions.

D. Effect of pH

1. Place 5 mL of 1% starch solution in each of 4 large test tubes. Add the following to the test tubes:

test tube	
1	5 mL pH 3 buffer
2	5 mL pH 5 buffer
3	5 mL pH 6 buffer
4	5 mL pH 9 buffer
5	5 mL pH 11 buffer

2. Place the tubes in a 37-40 °C water bath. After 5 minutes, add 5 drops of saliva to each of the test tubes as quickly as possible, mixing each solution thoroughly, and then returning the tubes to the water bath. *Do not overheat the bath or you will inactivate the enzyme.*

3. Prepare a spot plate for testing for the presence of starch in the samples by placing one drop of iodine reagent in each of five depressions on the spot plate.

4. Two minutes after the addition of saliva, transfer one drop from each test tube (using a different pipet for each tube) to a separate drop of iodine in the spot plate. Note the color produced for each.

5. Clean the spot plate and then prepare it for the next testing by placing one drop of iodine reagent in each of five depressions.

6. Repeat the testing at 5 minutes, and at 5 minute intervals thereafter. Continue testing for 30 minutes. In some tubes, the enzyme will be inactive, and the starch will always be present.

7. Record the time required for the hydrolysis of starch.

8. Plot the relative enzyme activity (time for starch to hydrolyze) versus the pH of the samples.

E. Inhibition of Enzyme Activity

1. Place 5 mL of 1% starch solution in each of 4 large test tubes. Add the following to the test tubes:

test tube	
1	1 mL 0.1 M $AgNO_3$ + 4 mL water
2	1 mL 0.1 M NaCl + 4 mL water
3	5 mL ethanol
4	5 mL water

2. Place the tubes in a 37-40 °C water bath. After 5 minutes, add 5 drops of saliva to each of the test tubes as quickly as possible, mixing each solution thoroughly, and then returning the tubes to the water bath. *Do not overheat the bath or you will inactivate the enzyme.*

3. Prepare a spot plate for testing for the presence of starch in the samples by placing one drop of iodine reagent in each of four depressions on the spot plate.

4. Five minutes after the addition of saliva, transfer one drop from each test tube (using a different pipet

for each tube) to a separate drop of iodine in the spot plate. Note the color produced for each.

5. Clean the spot plate and then prepare it for the next testing by placing one drop of iodine reagent in each of four depressions.

6. Repeat the testing at 5 minute intervals. Continue testing for 20 minutes. Record the time required for the hydrolysis of starch.

Report Sheet: Experiment 7
Hydrolysis of Starch by Salivary Amylase

Name ________________________________ *Date* __________ *Section* __________

Partner's Name __________________________ *Instructor's Initials* ________________

A. Effect of Enzyme Concentration

Test Tube	Drops of Saliva	Starting Time	Time Blue Color Fades	Time (minutes) for Starch Hydrolysis
1	1			
2	2			
3	5			
4	10			
5	20			

Graph: Enzyme Activity (Time for Starch Hydrolysis) versus Drops of Saliva

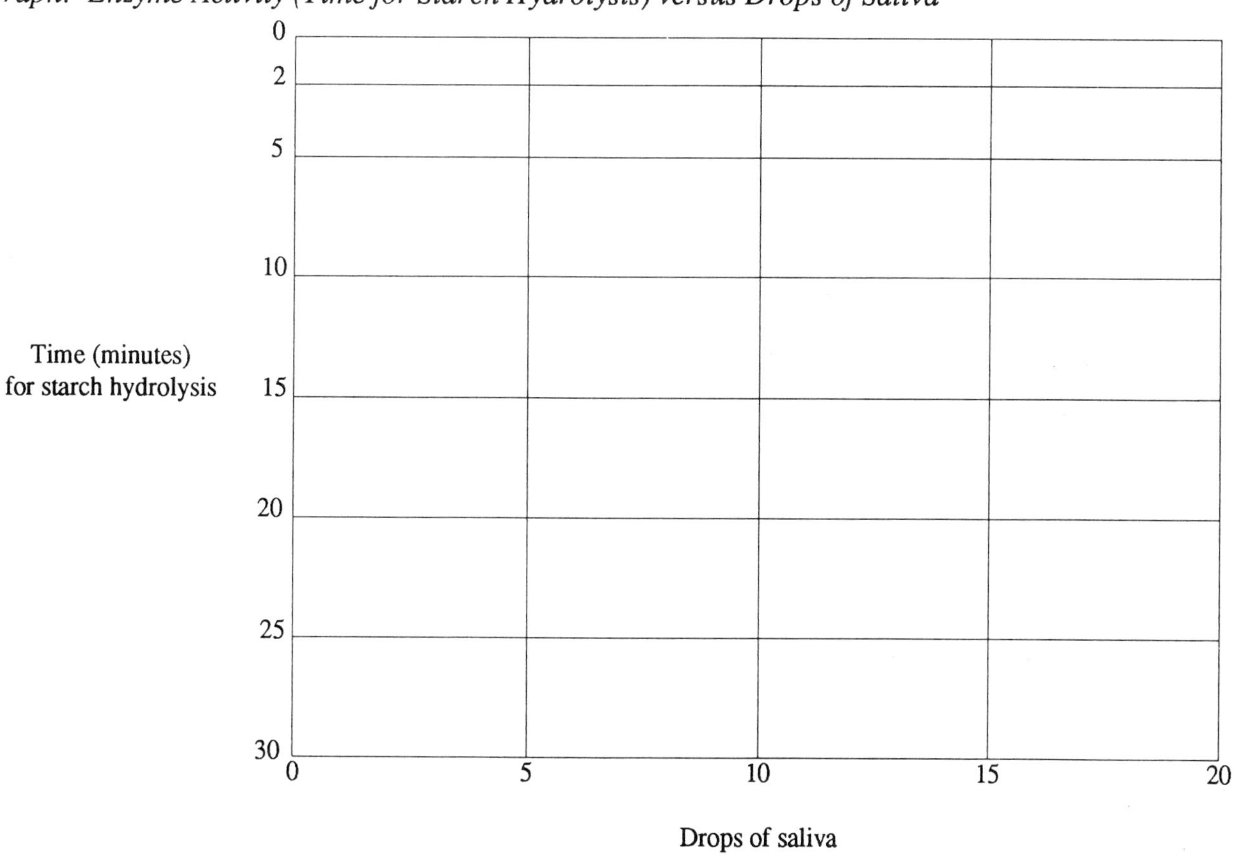

Questions:

1. How does the activity of the enzyme change as its concentration increases?

2. What are some reasons for this change in activity?

A. Benedict's Test for Easily Oxidized Groups

	Observations	Conclusions
Hydrolyzed Starch		
Fresh Starch		

B. Barfoed's Test for Monosaccharides

	Observations	Conclusions
Hydrolyzed Starch		
Fresh Starch		

Questions:

1. Are the results of the Benedict's and Barfoed's Tests what you had expected? Explain your answer by giving the structures and names of the products of complete enzymatic hydrolysis of starch.

D. Effect of pH

pH	Starting Time	Time Blue Color Fades	Time (minutes) for Starch Hydrolysis
3			
5			
6			
9			
11			

Graph: Enzyme Activity (Time for Starch Hydrolysis) versus pH

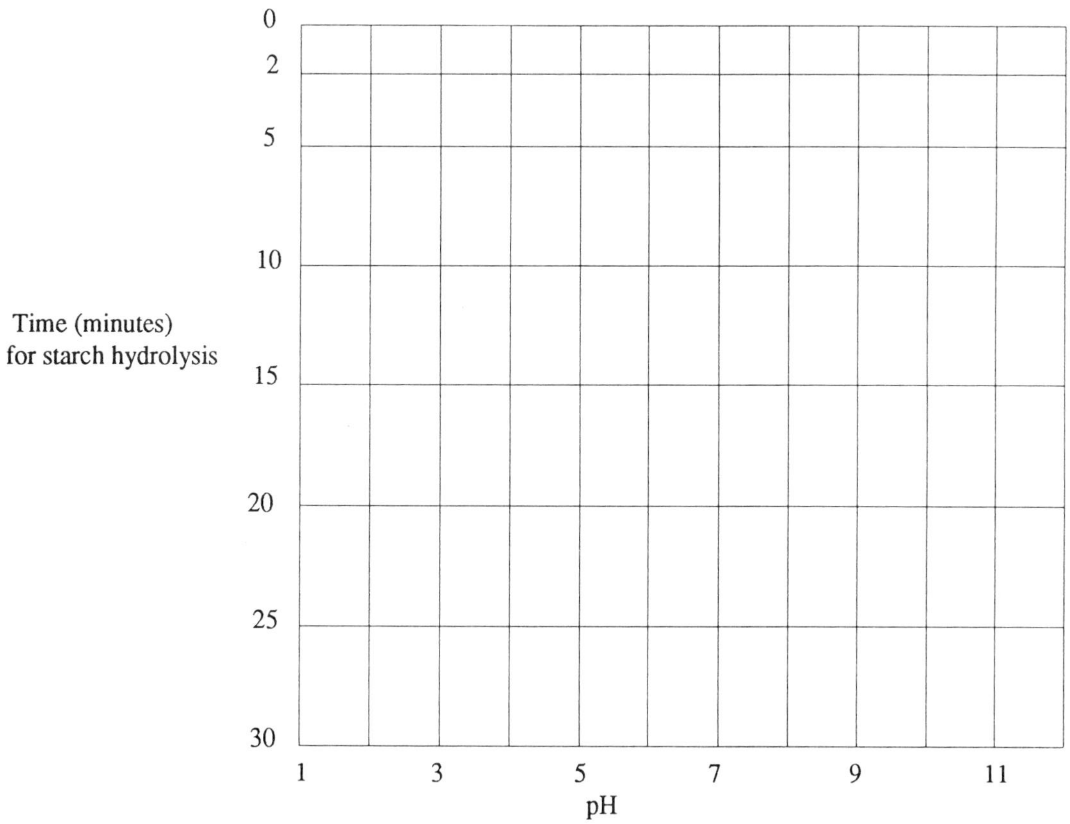

Questions:

1. What is the effect of pH on the activity of the enzyme?

2. What are some reasons for the effect of pH?

E. Inhibition of Enzyme Activity

Added Reagent	Starting Time	Time Bluc Color Fades	Time (minutes) for Starch Hydrolysis
$AgNO_3$			
NaCl			
ethanol			
water			

Questions:

1. Which of the compounds added to the reaction tubes are inhibit enzyzme activity?

2. Explain how the compounds you named above inhibit enzyme activity.

Additional Question:

Cellulose is the most abundant polysaccharide in nature; it is the structural material in plants. Is cellulose of nutritional value to man? Explain why or why not.

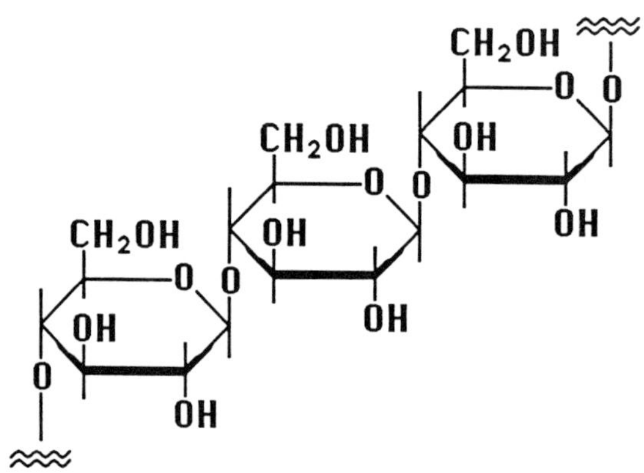

Experiment 8

Analysis of the Vitamin C Content of Fruit Juices

What you will accomplish in this experiment

You will perform an iodometric titration to determine the concentration of vitamin C that is present in a commercial fruit juice. This concentration will be expressed as milligrams of vitamin C per milliliter of fruit juice.

Concepts to know to be prepared

Ascorbic acid, more commonly known as vitamin C, is an essential nutrient. The compound has an important role in biological oxidation-reduction reactions and aids in the repair of damaged tissues. A vitamin C deficiency in humans will lead to scurvy, a condition that is characterized by sore gums and loosening of teeth, as well as anemia, edema, pain in the joints, and hemorrhaging. Vitamin C cannot be synthesized in the body, so it must be obtained from the diet. Citrus fruit, berries, broccoli, cabbage, sweet peppers, tomatoes, cabbage, and turnip greens are among the best food sources of vitamin C.

The quantity of vitamin C in a food source can be determined by titration with a tri-iodide (I_3^-) solution, which is prepared by combining iodine (I_2) with iodide (I^-). Ascorbic acid is oxidized by the I_2 (as I_3^-) to an inactive form.

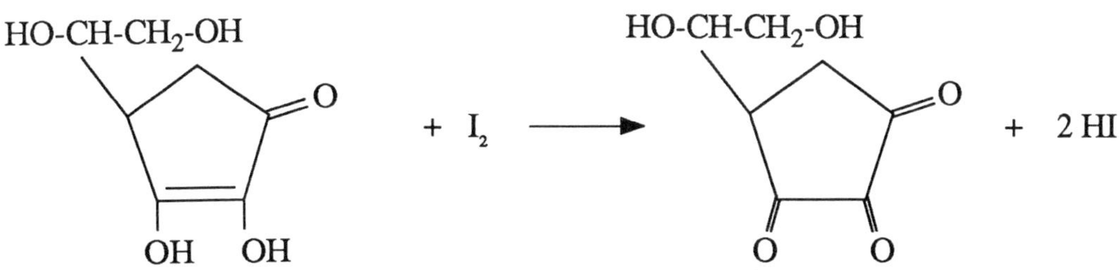

ascorbic acid (vitamin C)
formula weight = 176.13 g/mol

oxidized product

As the ascorbic acid is oxidized by I_2, the I_2 is reduced to I^-. When all of the ascorbic acid has been oxidized, any excess I_2 will react with a starch indicator to form a bluish-black colored starch-iodine complex. The formation of this color will signal the end-point of the titration.

The following example will illustrate how to determine the weight/volume percent of vitamin C in fruit juice from titration data:

5.73 mL of a 0.01005 M iodine solution are required to titrate the vitamin C in a 30.00 mL sample of fruit juice. What is the concentration of vitamin C in the fruit juice (expressed in milligrams vitamin C per milliliter of fruit juice)?

$$\text{moles of } I_2 \text{ required for titration} = 5.73\text{ mL} \times \frac{1\text{ liter}}{1000\text{ mL}} \times \frac{0.01005\text{ mole}}{1\text{ liter}} = 5.76 \times 10^{-5}\text{ moles } I_2$$

$$\text{moles vitamin C titrated} = 5.76 \times 10^{-5}\text{ mole } I_2 \times \frac{1\text{ mole vitamin C}}{1\text{ mole } I_2} = 5.76 \times 10^{-5}\text{ mole vitamin C}$$

$$\text{mass of vitamin C titrated} = 5.76 \times 10^{-5}\text{ mole vitamin C} \times \frac{176.13\text{ g}}{1\text{ mole}} = 0.0101\text{ g vitamin C}$$

$$\text{concentration of vitamin C in fruit juice (mg/mL)} = \frac{0.0101\text{ g}}{30.00\text{ mL}} \times \frac{1000\text{ mg}}{1\text{ gram}} = 0.337\text{ mg/mL}$$

Procedure that you will follow

A. Visualization of the End Point Color

1. Place about 50 mL of distilled water into a small beaker. Add 20 drops of starch indicator (a 2% starch solution), then three to five drops of the iodine solution. Carefully observe the blue-black color that is produced. This is the color change that will signal the end point of the iodometric titration of the fruit juice.

B. Iodometric Titration of Fruit Juice

1. Record the brand name of your fruit juice. Pour about 60 mL of the juice into a clean, dry, 100-mL beaker. If the fruit juice is cloudy or contains particulate matter, it can be clarified by adding a small amount of Celite (a filter aid), mixing throughly, and pouring the mixture through a funnel plugged with a large piece of cotton. The filtrate can be collected in a small Erlenmeyer flask; the mouth of the flask will support the funnel.

2. Use a 10-mL volumetric pipet (or buret or graduated cylinder) to transfer 10.00 mL of the clarified fruit juice into a 125-mL Erlenmeyer flack. Add 20 mL of distilled water, five drops of 3 M HCl (a catalyst), and 20 drops of starch indicator.

3. Prepare a 50-mL buret for use by washing it with a mild detergent (if necessary), then rinsing it first with tap water, then with distilled water, and finally with iodine solution. Fill the buret with the iodine solution, and adjust the meniscus to at or below the zero mark. Read the initial buret volume (to two places past the decimal) and record the value. (Care and use of the buret is described on pages 7 and 8 of this laboratory manual.) Record the molarity of the standard iodine solution that is provided to you.

4. Place the flask containing the vitamin C mixture under the buret and add iodine solution dropwise, with swirling, until the color of the solution *just* changes to dark bluish-black. This color should persist for at least 30 seconds. Read and record the final buret volume.

5. Repeat the titration procedure two additional times, using 20.00 and 30.00 mL portions of the fruit juice.

6. For each trial, determine the total volume of iodine required for the titration, and determine the concentration of vitamin C in the fruit juice in mg/mL.

Report Sheet: Experiment 8
Analysis of the Vitamin C Content of Fruit Juices

Name ______________________ *Date* __________ *Section* __________

Partner's Name ______________________ *Instructor's Initials* __________

Brand of fruit juice ______________________

Molarity of standard iodine solution ______________________

A. *Iodometric Titration Results*

	Sample 1	Sample 2	Sample 3
Volume of fruit juice used	________	________	________
Initial buret reading	________	________	________
Final buret reading	________	________	________
Volume of iodine solution used	________	________	________
Moles of iodine used	________	________	________

Calculation for sample 1:

	Sample 1	Sample 2	Sample 3
Moles of vitamin C titrated	________	________	________

Calculation for sample 1:

	Sample 1	Sample 2	Sample 3
Mass of vitamin C titrated	________	________	________

Calculation for sample 1:

	Sample 1	Sample 2	Sample 3
Concentration of vitamin C in fruit juice (mg/mL)	________	________	________

Calculation for sample 1:

Average concentration of vitamin C in fruit juice ____________________

Calculation:

Additional Question

According to your average above, how many ounces of fruit juice should you drink in order to consume 60.0 mg of vitamin C? 60 mg is the recommended daily allowance (RDA) for vitamin C set by the National Research Council. (33.81 fluid ounce = 1 liter)